# Management of Change in Water Companies

# Management of Change in Water Companies

*In Search of Sustainability and Excellence*

Joaquim Poças Martins

IWA PUBLISHING

**Published by** **IWA Publishing**
**Alliance House**
**12 Caxton Street**
**London SW1H 0QS, UK**
Telephone: +44 (0)20 7654 5500
Fax: +44 (0)20 7654 5555
Email: publications@iwap.co.uk
Web: www.iwapublishing.com

First published 2014


*British Library Cataloguing in Publication Data*
A CIP catalogue record for this book is available from the British Library

ISBN: 9781843391951 (Paperback)
ISBN: 9781780406510 (eBook)

To the future of Beatriz, Benedita, Alice, Luísa and Manel.

# Contents

**Chapter 3**

# About the Author

**Joaquim Poças Martins**

During the last 18 years Joaquim Poças Martins has been the CEO of the largest Portuguese water companies, namely *Águas de Portugal, EPAL (Lisbon), Águas do Douro e Paiva* and *Águas de Gaia,* and as a consultant he has coordinated the restructuring processes of *Águas do Porto* and other water utilities.

He has a degree in Civil Engineering from the University of Porto, Portugal, an MSc in Public Health Engineering and a PhD from the University of Newcastle upon Tyne, UK, and a Diploma in Top Management from IESF, University of Navarra, Spain.

He is a fellow of the Portuguese Institution of Chartered Engineers and member of the IWA Water Loss Specialist Group.

He is Professor of Hydraulics at the Faculty of Engineering of the University of Porto, the Secretary General of the Portuguese National Water Council and Chairman of the company *GIDEA – Innovation and Development for the Water Industry*.

Between 1993 and 1995 he was the Secretary of State for the Environment and the Consumer in the Portuguese Government.

He is the author of two books and of over one hundred publications.

## Co-Authors

**Ricardo Simões Campos**
Co-Author of the Copasa Case Study

Ricardo Simões Campos graduated in Civil Engineering from the Faculty of Engineering of the Federal University of Juiz de Fora, with a specialization in Sanitary Engineering from the Federal University of Minas Gerais, Brazil.

He joined Copasa in 1977, where he developed a wide range of managerial functions: Secretary General to the Presidency, Planning and Control, and Operational Control of the Metropolitan Area.

He was elected to the Board in 2004 as CFO and in 2009 became the CEO and President of the Company.

**Cassilda Teixeira de Carvalho**
Co-Author of the PNQS Case Study

Cassilda Teixeira de Carvalho has been Copasa's President Technical Assessor since March 2003. She has worked for *Companhia de Saneamento de Minas Gerais* since November 1979 and executed technical functions in the areas of water and wastewater systems design, construction and operation, and also in the planning and implementation of water and wastewater systems for rural areas. She has also been in charge of institutional development and business performance evaluation at the company.

She is a former President of the Brazilian Association of Sanitary and Environmental Engineering – ABES – the main association in the sector in Latin America, with over 10,000 affiliates in Brazil, from 2008 to 2012. She is also on the Board of Directors of the International Water Association – IWA – as a Latin America representative.

She graduated in Civil Engineering at the Federal University of Minas Gerais in 1979 and earned her master's degree in Sanitary, Environmental and Water Resources Engineering at the same university.

# Foreword

This is a book about change – and managing change – in water utilities. Challenging existing attitudes of water utility managers, changing their thought processes and increasing their level of awareness is essential for improving a utility's operational and financial efficiency. Putting in place strong leadership is usually a prerequisite to making these changes. The performance indicator that is widely accepted as a major indicator of a utility's overall operational and financial efficiency is the volume of non revenue water (NRW). This is the volume of water lost from the pipe network and the loss of income to the utility from inadequate customer metering and billing. However, water utility managers often lack awareness of the significance and value of reducing NRW – which effectively provides another source of water. Utility managers are tempted to take the easier route to meet increasing customer demand by augmenting water treatment works and upgrading pumping stations. The first step to increasing a utility's revenue stream is often to increase tariffs, rather than maximising income from customers by putting in place better and more efficient customer metering technologies and revenue collection policies.

Joaquim Pocas Martins is well placed to write about managing change to improve efficiency. His post as Secretary of State for the Environment gave him valuable experience in understanding the Portuguese water sector, the attitudes of utility managers and the need for restructuring their utilities. This experience was consolidated in later years in his role of 'turning round' water utilities that were 'flat-lining' in their attitudes to effective utility management and NRW management.

The book illustrates the challenges and successes of NRW reduction and energy savings in several case study examples – with some outstanding results. One study shows how, over a six year period, water losses were more than halved, producing sufficient extra revenue to invest in new capital works with no increase in charges to the customer.

There are still many water utilities around the world that do not fully understand – nor appreciate – the need to change attitudes, and sometimes management structures, to provide the leadership and skills required to make their utilities more efficient.

That is why this book is very timely. Joaquim Pocas Martins' skills in people management, and his attitude to technical innovation, provide an excellent knowledge base and valuable experience for this book. The case study examples provide a useful tool for any water utility manager, and for water utilities, striving to change the structure and attitudes of their companies to NRW.

I am pleased to recommend this book to them.

Malcolm Farley<br>Deputy Chair/Secretary<br>International Water Association<br>Water Loss Specialist Group

# Preface

Many water utilities around the world are being managed with surprisingly low levels of efficiency. Some of these utilities are run with chronic losses or are profitable only because they practice excessively high tariffs, receive lost fund transfers or are continuously subsidised with tax payers' money from municipal, regional or national budgets.

Due to this generalized state of inefficiency and to the difficulty of the public sector to finance the construction, rehabilitation and expansion of infrastructure in the past decades, there have been privatization and nationalization waves and restructuring processes with mixed results.

It is known that when inefficient water companies start to be well managed, by public or private entities, their performance indicators improve very fast.

There is a great difference between managing continuity and managing change. This book is focused on the process of change that is required to transform poorly performing utilities into efficient and sustainable ones. The methodology presented here is innovative in many aspects and has been successfully implemented in the change management processes of real water companies. Restructuring a water utility is a very complex process, which requires political support, leadership, financial resources, time and an adequate context. It is argued that usually a four to six year transition period is necessary to restructure a water utility; however, it is important to take advantage of every window of opportunity and it is possible to obtain great results in a short period of time, especially in poorly performing utilities.

Reducing Non Revenue Water (NRW) and other inefficiencies is essential to succeed and a new computer program prepared to be used at the top management level is presented in the book; it optimizes the whole NRW reduction process, leading to the Economic Level of Loss, which may be defined in different forms.

A simplified restructuring model is also presented in the book, based on four critical points: (1) definition of the sustainability boundaries of the water utilities, (2) reducing NRW and other inefficiencies, (3) investment for universal coverage and finally (4) adjusting tariffs.

The methodologies presented in the book are illustrated through five case studies: three from Portugal and two from Brazil.

The restructuring of a poorly performing water utility is essentially a transition phase in which it is desirable to move fast and with determination in search of excellence. Like any transition period, it must come to an end, and it should be followed by a period of stability, maintaining or improving on what has been achieved.

This book is essentially about the restructuring process but it also takes into consideration the following stability period. In fact, it would not be correct simply setting short term targets, as a water utility's major assets are expected to last for many decades and the need for water services is eternal. The basic methodologies for restructuring water utilities presented in the book may be used under different management and ownership models. The problems to be solved in poorly performing utilities are similar but the solutions are site specific, and there is often space for imaginative, unconventional solutions, as shown in some of the examples presented in the book.

# Acknowledgements

The following tables and figures are used in the book, under the authorization of their authors or publishers:

- Figures 2.3 and 2.7 (OECD)
- Figure 2.9 (ERSAR)
- Figure 3.1 and Table 3.2 (World Bank)
- Figure 3.11 (Malcolm Farley and Stuart Trow, IWA Publishing)
- Figure 3.13 (Alan Lambert)
- Figure 4.5 (Pinsent Masons)

The case studies on *Águas do Porto* (2006-1013), *Águas de Gaia* (1997-2004) and *Águas de Portugal* (2004-2005) refer to restructuring processes led by the author during the respective periods, and were based on publicly available information: the corporate annual reports and public presentations given by the author on behalf of the companies. The case study on Copasa is co-authored with Ricardo Simões Campos, the CEO of the company. The case study on the ABES PNQS is co-authored with Cassilda Teixeira de Carvalho, former President of ABES and President of the PNQS.

The author wants to thank the collaborators of *Águas de Gaia, Águas de Portugal, EPAL (Lisbon), Águas do Douro e Paiva and Águas do Porto* for their knowledge, contributions and commitment to the results of the successful change management processes that were at the basis of the decision to write this book.

This book would not exist also without the initial encouragement and support of Malcolm Farley, a widely acknowledged international water loss consultant and author, who also kindly agreed to write the Foreword. Paula Malheiro collaborated in the recording, formatting and research for the earlier version of the book. My colleague and friend Andrew Donnelly gave very important suggestions on the contents of the book and helped in the final revision.

My son Ivo, a very versatile architect, created most of the original figures and my son João Pedro, also a Professor at the University of Porto, gave valuable suggestions and advice on the content and helped in the final revision of the book. I thank my closer family and friends for their support, encouragement and patience, especially during the final phase of intensive writing.

Finally, Idedicate this book to the billions of people in the world who may be paying too much for their water services due to inefficiencies that could be avoided.

# Chapter 1
# Introduction

Tolstoi, in Ana Karenina, says: "happy families are all alike, every unhappy family is unhappy in its own way."

Similarly, all good water utilities look alike: sustainable, efficient, low levels of water loss, with satisfied clients, shareholders, employees and regulator, whilst poor performing ones have many different excuses in order to explain to their customers, for example, their unacceptable levels of water loss or high tariffs.

Improvement requires change, and this may take place at a slow or fast pace and assume many different forms. Unlike other organizations subject to market forces, in the absence of competition, many water utilities have managed to avoid change and survive over long periods of time providing a poor service or an expensive service, or often both. This is becoming less and less true, as customers and the media put increasing pressure on the management of utilities: supply interruptions and failure to meet water quality standards are almost always publicized, and tariffs are discussed and compared.

Because change is becoming more frequent and important, continuity management is just not enough and must be accompanied by leadership, which is the primary force behind successful change. This book is based on real experiences, with successes and failures, of water utilities that went through processes of change. It reflects the personal experience of the author from leading processes of change in different water supply/sewage companies, and it also includes examples of very large-scale change management processes in Brazil.

The book is mainly about the restructuring of water utilities and how to:

- modernize old-fashioned organizations;
- make companies profitable;
- promote efficiency;
- deal with overstaffing;
- plan, finance, build and maintain infrastructure;
- introduce innovation;
- deal with customers, owners, the regulator, the workers, the unions and the media; and
- achieve sustainability.

It is intended for people involved in the management of not just water utilities, as the concepts and theories to promote change in organizations are common to other public and private organizations.

Water loss, usually considered only from the obviously negative point of view, is taken here also as an opportunity for improvement and it is suggested that change management should go hand in hand with the water loss reduction process.

The change management methodology presented in this book starts with a Water Balance - using the IWA Water Loss Task Specialist Group approach - and with the positioning of the water utility in a four-stage classification system, according to its current level of development: pre-efficacy, efficacy, efficiency and excellence. Excellence is related to the Economic Level of Loss (ELL) and to the 3T sustainability model (Tariffs, Transfers and Taxes). A new computer program has been developed to support this change management methodology. Using around fifty site-specific figures that can largely be obtained from water utility annual reports, this program builds the initial water balance and, based on the user specified site-specific water utility data, it calculates the seven different components of the Non Revenue Water (NRW). These components can be regarded as seven "diseases" with a common symptom: NRW. The therapy for these "diseases" is based on four groups of measures: basic, priority, medium-term and long-term measures. Unlike other approaches, in which the ELL is estimated *a priori* or calculated on a single set of assumptions, this computer program defines it as a result of a formal optimization process that takes into consideration the impact, costs and benefits of each measure to reduce the above mentioned seven NRW components. Designed for top management of water utilities and for the set of variables and criteria selected by the user, the program automatically calculates the ELL, the necessary investment plan, the measures to be taken and the corresponding expected results in terms of NRW reduction, including the final water balance.

The book also addresses the main aspects of restructuring water utilities and presents a simplified restructuring model based on the definition of a four to six year transition period during which four major actions should be taken: (1) redefinition of sustainability boundaries of the water utility; (2) improving efficiency and reducing NRW; (3) investing in infrastructure to achieve universal coverage and reliability and finally; (4) adjusting tariffs to achieve full cost recovery.

Two case studies presented in this book show that the savings from reducing water loss and other inefficiencies may be sufficient to pay for the whole change management process, including rehabilitation investment and system expansion as well as expansion of activities to include the integrated management of all components of the urban water cycle (water supply, wastewater, stormwater, urban streams and beaches), without tariff increases.

Excellence needs to be actively maintained; otherwise the situation can easily regress to where it was before. Using the water utility development stages presented in the book, a post-excellence stage, therefore, could well be a return to Pre-Efficacy, for example if some drastically wrong political decisions concerning the future of the water utility are made. To avoid this situation, it is necessary to make sure that, after the restructuring process, there is a sufficiently long stability period, taking measures actively to keep and improve the achieved values of the performance indicators, principally NRW. This is the time for maintaining Active Leakage Control and management and to develop programmed investment plans for asset rehabilitation and replacement.

This book is organized in five chapters, following the Introduction.

*Chapter 2* – Change Management, Sustainability and Non-Revenue Water-summarizes the major concepts of the book and shows how it is possible to achieve sustainability through a change management process in water utilities using the savings from reducing Non Revenue Water and other efficiency gains.

*Chapter 3* – The Efficiency Challenge-deals with the identification of the major factors that determine the efficiency of water utilities, especially Non-Revenue Water. A new interactive optimization program for NRW reduction, designed to be used for target setting purposes at the top management level, is outlined.

Based on a small number of water utility key indicators, it automatically calculates the ELL based on different assumptions and presents a strategy to achieve it, including the individual actions and the corresponding savings in terms of volume of water and money.
*Chapter 4* – Restructuring Water Companies-presents a number of especial issues and recommendations regarding the restructuring of water utilities derived from the case studies presented and from the author's experience in change management processes.

*Chapter 5* presents five case studies:

*Case Study 1*

- looks at the change management process which took place under public management at the municipal water utility *Águas do Porto*, in Portugal, during the period 2006-2013. It shows how it was possible to halve Non-Revenue Water in just eight months, to reduce it by 80% over seven years and to reduce energy consumption used for water pumping by more than 95% in five years. It also shows the great advantages of integration of the urban water cycle in terms of the rehabilitation of urban streams and beaches.

*Case Study 2*

- looks at the restructuring process of another Portuguese municipal water utility, *Águas de Gaia*, between 1997 and 2004, in which the most interesting aspects were the construction, in six years, of a previously almost inexistent wastewater system and the integrated management of the urban water cycle, mostly financed through NRW reduction.

*Case Study 3*

- features COPASA from *Minas Gerais*, Brazil, one of the largest water utilities in the world. It is publicly managed, state owned and listed on the Stock Exchange. The most relevant aspect of this Case Study is the whole change management process, which started in 2003 and was referred to locally as "shock management". It transformed the water utility into a worldwide success story with very interesting examples of social projects for the low income population without affecting the sustainability and profitability of the company and its tariffs.

*Case Study 4*

- concerns the PNQS - National Plan for Quality in Sanitation-from the Brazilian Association of Sanitary Engineers (ABES), one of the largest water and sanitation professional associations in the world, with over 10,000 members. PNQS gives awards to the best water and sanitation projects in the country, but its importance is far greater than the prizes that everybody in the Brazilian water and sanitation sector would be proud to receive: the motivation to apply, the training required to do so, the team work (from top to bottom, or vice versa) and the knowledge that is acquired in the process have been responsible for an important part of the improvement that has taken place in one of the countries that has been investing most in the water sector worldwide.

*Case Study 5*

- refers to the Portuguese holding *Águas de Portugal* during the short 2004-2005 period. This is an example of a very ambitious restructuring process at the national level which, at the time, had very ambitious but not sufficiently clearly defined political objectives, and that was aborted at an early stage due to an unexpected change in the political cycle. It is presented here to illustrate the importance of having clearly defined objectives at an early stage, the need for political support in the whole change management process and the complexity of restructuring a water utility sector with many small municipalities.

# Chapter 2

# Change Management, Sustainability and Non-Revenue Water

## 2.1 INTRODUCTION

Surprisingly, discovering that a water company has a high level of water that is not converted into revenue, namely NRW may be a good sign. In fact, if the water company has managed to survive and deliver a service under such circumstances, it has, at least, a new source of additional income to explore. The savings from reducing NRW may be enough not only to pay for the control and reduction process, but also to finance restructuring of the water company, including rehabilitation, investment and expansion of the system. Indeed, the reduction of NRW is a common goal of most water companies around the world, large or small, public or private, in developed or developing countries. Besides technical knowledge and sophisticated technology, its reduction requires the commitment of top management and organizational change.

NRW, together with other indicators, helps to position the development stage of each water utility in a pre-efficacy/efficacy/efficiency/excellence rating scale, as presented in this book. The process of evolution along this rating scale is a common pattern for water companies in search of excellence (Pocas-Martins, 2009a and b). Based on case studies, that do not include situations of generalised poverty or non-payment of water bills, it is argued that five to six years are sufficient, under ideal conditions, to approach excellence, but also that every window of opportunity to improve on the existing situation must be explored. In previously less efficient companies, it is possible that a tipping point, the level at which the momentum for change becomes unstoppable, may occur in the first year, with the creation of tangible results. Each water company has its uniquely complex context, but political support, leadership and time are prerequisites for success whilst excellent water utilities are, by definition, economically, environmentally and socially sustainable.

The first level of sustainability is related to the economic component, which is based on the '3T' model of Tariffs, Transfers and Taxes (OECD, 2009). This chapter shows how to reach sustainability and excellence in a water utility through a change management process, with the reduction of NRW and other inefficiencies.

## 2.2 CHANGE MANAGEMENT IN WATER UTILITIES

### 2.2.1 The Chain of Change

Unlike other organizations that are strictly subject to market forces and in the absence of competition, many water companies all over the world have managed to avoid change and survive over long periods of time while providing a poor and/or expensive service. This situation is becoming less and less acceptable, as more stringent water legislation is enforced whilst consumers, regulators and the media are exercising increasing pressure on water company performance. Interruptions of supply and failure to meet water quality standards are less and less tolerated, whilst performance indicators and tariffs are increasingly compared through benchmarking between utilities, penalizing the poorly performing ones. Change, therefore, is increasingly becoming the rule in water companies, whether it is externally driven or simply the result of a strategic management decision.

The depth of change required depends primarily on the level of development and management of each water utility: some will require only improvements resulting from knowledge and technology, others will also need alterations to the internal organization and middle management whilst, finally, others are in such a poor situation that they require restructuring or a change management process (Figure 2.1).

**Figure 2.1** Depth of change required.

Independently of the depth of the change required in a certain water utility, at a given moment, what must be done depends on a number of conditions, some of which are general, some site-specific. It is possible to establish an analogy between the change processes and the use of a chain to pull a heavy weight. The chain strength is famously limited by the strength of its weakest link, therefore to increase that strength it is necessary to identify and reinforce the weakest link. The strength of this reinforced chain becomes limited by the former second-weakest link; this is the next to be reinforced, and so on, until the pulling force and not the strength of the link becomes the limiting factor.

To adapt this general analogy to a change management process, it is necessary to identify the individual factors that are critical to success. From personal experience, these are, in decreasing order: political support, leadership, people, money, time and context (Figure 2.2).

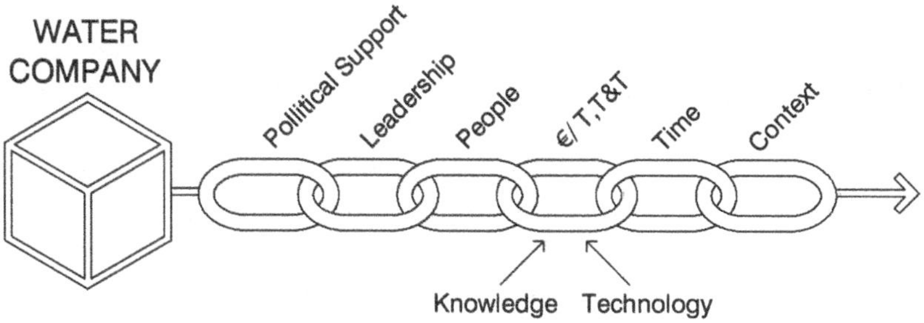


**Figure 2.2** The chain link analogy for the change management process in water utilities. (TTT = Tariffs, Transfers and Taxes; see below).

#### *2.2.1.1 Political support*

The majority of water utilities are publicly owned and managed, but even when they are privately managed, they are municipal or state monopolies and the top decision maker is still a politician. It is therefore virtually impossible to start a change management process without political decision and support. Processes of this type tend to take place at the beginning of political mandates, triggered by public complaints about poor quality of service or about tariffs and charges. From personal experience, this is by far the most important success factor: even the most competent utility leader, without political support, can do very little.

#### *2.2.1.2 Leadership*

The second link in the chain is related to leadership. Once a political decision for change has been made, it is up to the mandated politician to appoint the project leader. The depth of this process determines the type of leader required. In fact, most of the successful change management processes reported have one thing in common: the appointment of a change agent, a leader who was given both political support and conditions to succeed. Traditional management is just not sufficient or adequate for change: leadership is really the primary force behind successful change. The ideal leader for a change management process in a water utility is someone who has been granted political support, has leadership skills and who has had previous experience in restructuring water companies. Of course, it is not always easy to find someone with these characteristics and inevitably some compromises have to be accepted. In particular, as change management is ephemeral and the prelude to future stability, it is possible to contract the specialized services of an external consultant to closely assist the management board in charge of the process. From personal experience, this solution has a good opportunity to succeed if there is a strong relation of trust between the board and the consultant.

Leadership and management are different and complementary concepts: leadership without management leads to organizational chaos, management without leadership is bureaucracy; at times of change, leadership is definitely paramount. It is interesting to acknowledge that charisma, another face of leadership, can be uncovered when reaching power, which explains, for instance, why the bosses' jokes tend to be funnier.

There are many books and articles written about leadership and this is not the place to innovate on this subject. It is sufficient to acknowledge that there are no universal rules to becoming a good leader, just as there are no rules for being a good parent or a good teacher. Leadership is, to a large extent, instinctive, but can be improved through practice, study and age.

However, leadership should not be confused with leadership style the most effective leaders can act and switch between various styles depending on the situation. Leadership is particularly important in the initial stages of a change management process, when it is necessary to take many important decisions at short notice, with limited information, without knowing key personnel well and maintaining consideration of the great risks of interrupting or deteriorating service quality.

#### *2.2.1.3 People*

It is often said that people are the most valuable resource of organizations. This is not precisely correct: in fact, good workers and those who are willing to be good are, indeed, the most important resources of organizations. Even the most talented leader cannot change an organization alone: every decision he or she makes has to be interpreted and implemented through or at least with other persons; therefore, identification and selection of key collaborators is essential.

Any organization relies on its own employees and on those of external contractors. There are performance indicators that refer to the ideal number of workers per number of customer connections or per unit of length of pipes, for instance, including both the internal and external staff. In the past, there was hardly any use of outsourcing in water utilities and almost every task was undertaken internally by employees. The education system did not feed the job market with people that water companies could use straight away for many of their specialized activities, making it necessary to develop internal apprentice training. It is often argued that core business activities must be undertaken internally and should not be outsourced; however, conditions are changing very fast. In fact, it depends on what the market has to offer in any particular location. In the developed world there is an increasing use of outsourcing for almost every specific service and the educational system produces skilled people for virtually every activity. However, in many developing countries it is very difficult to find trained people in the job market and therefore this "link" is often found to be the weakest and the most difficult and time consuming to reinforce.

#### *2.2.1.4 Money*

In any water utility, money is essential to finance the capital and operational expenditure. According to the previously mentioned 3T economic sustainability model (detailed below) (OECD, 2009), there are three sources of financing: tariffs, transfers and taxes.

The only long term sustainable source of funds for water utilities is customer tariffs, but most water utilities in almost every developed or developing country have received, at some stage, transfers in the form of taxpayer contributions, in order to avoid rising tariffs to unaffordable or socially unacceptable levels, as shown in Figure 2.3.

Moreover, there are some situations, such as in remote areas or on small islands, where it is necessary to subsidize not only the construction costs but also operational costs. Furthermore, it is often necessary to support expansion and provision of water services to the poorest groups of the population.

Knowledge and technology are absolutely essentially to the success of the change management process. In some cases, especially in developing countries, even if there is money available, knowledge and technology may be difficult to obtain but, in general, both can be acquired in the market by water utilities: by contracting consultants and services and purchasing or renting equipment. Consultants and technology are necessary, but, on their own, they do not create sufficient conditions to produce a significant improvement in a water utility. Indeed, they are only effective if there are people in the organization who are motivated and trained to absorb that knowledge and to make good use of new technologies. There are many examples in water utilities of useless consultant missions and of expensive equipment for leak detection or video monitoring that have never been used.

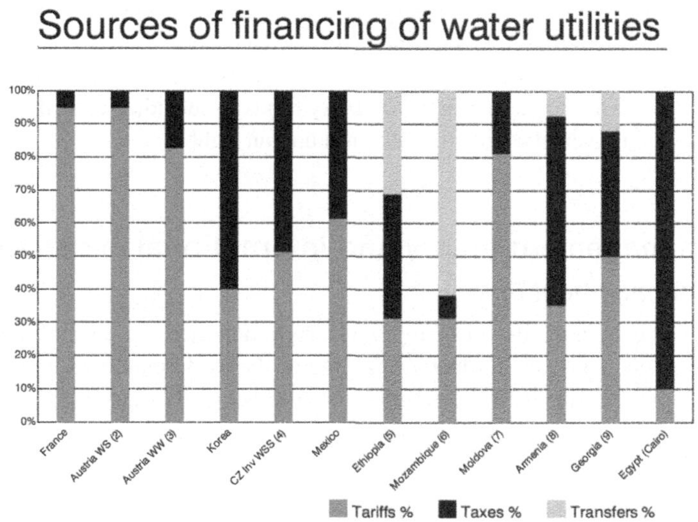


**Figure 2.3** Sources of financing of water utilities in different countries (Source: OECD, 2009).

### *2.2.1.5 Time*

The case studies presented in this book show that it is possible to fundamentally change water companies from poor performance to excellence in half a dozen years. This period of time, in ideal conditions, is sufficient to restructure an organization, to plan, design and build complex treatment plants and to install or replace hundreds or even thousands of kilometres of pipes. However, there are impediments of a non-technical nature, such as environmental, legal, financial and political issues, that may introduce delays or even abort the process.

In publicly owned water companies, management tends to follow or at least be affected by political cycles. This may limit the possibility of developing and implementing medium- and long-term strategies and change management processes. However, especially in municipalities, the political cycles tend to be longer, lasting for a number of consecutive mandates of often a decade or more; on the other hand, the participation of private shareholders in public water utilities usually leads to the appointment of professional managers, less affected by the political cycles. When a management team has only one or two years of continuity and stability ahead, it does not make sense to start a deep restructuring process, but it is always possible to introduce measures to improve the organization and efficiency of the water utility. It is the leader's responsibility to incorporate the existing time limitations into the objectives of the corporate strategy. Errors of judgement in this respect may be fatal: it is always better achieving 10% of something than 100% of nothing.

### *2.2.1.6 Context*

The context in which a water company operates is crucial to the process of change. In cases of war zones or of generalized chaos or poverty, it may be very difficult to improve a water utility, or indeed any other organization. If a municipality, region or country is not stable, there may not be conditions in which to tackle the restructuring of a water utility.

Where poverty prevails, it is very important to set goals which are aligned with the availability of investment subsidies (transfers) and the customers' ability and willingness to pay. Even under this difficult context, it may be possible to design appropriate solutions based on the 3T economic sustainability model.

These solutions are in principle bankable, which is, in itself, an added value. Otherwise, if the costs cannot be covered by a combination of tariffs, transfers and taxes, we face a problem that requires regional, national or international solidarity. However, if the number of poor customers is very high in relation to the paying ones, it may be impossible to mobilize sufficient solidarity funds to subsidize even the social tariffs, leading to an effective denial of the universal access to water as a human right.

## 2.2.2 Change management theory and international experiences

### *2.2.2.1 Change management theory*

Change management has become a discipline in its own right and there are many books and best practice manuals on how to manage change in organizations. One of the leading authors in this field, John Kotter, recommends that, in a change management process, the following steps are followed, in this exact order (Kotter, 1996; Jones, 2010):

(1) Create a sense of urgency;
(2) Put together a strong enough team to direct the process;
(3) Create an appropriate vision;
(4) Communicate the new vision broadly;
(5) Empower employees to act on the vision;
(6) Produce short-term results;
(7) Build momentum and use that momentum to tackle the tougher change problems;
(8) Anchor the new behaviour in the organizational structure.

Kotter notes that there is ample evidence that these steps were followed in successful change processes, whilst in other cases, missing some of them or changing their order of implementation has often resulted in failure. The author can confirm, through personal experience in two case studies presented in this book, that this theoretical model is suitable to the restructuring of water utilities.

The sense of urgency that triggers the change process within a water utility may be a political option or, for instance, a generalized public complaint about quality of service or high tariffs.

The selection of the project leader is the first condition to overcome the next stage on the list. Besides leadership qualities, he or she should have previous experience in change management projects and preferably in the management of water utilities. The possibility of contracting good professionals and rewarding the best collaborators are essential to create a strong team to effectively direct the project.

The new corporate vision should be clear and mobilizing, for example "becoming a world class water company" or "becoming the country's best water company". However, less ambitious statements may be adequate if the context is not favourable and, in any case, it is better that the project leader refines the definition of the vision with the participation of the closer collaborators. Communication of the vision to utility staff at all levels contributes to the alignment of individual and corporative objectives, contributing to a higher level of employee performance and having a multiplier effect on the whole organization. As is clear in the water company case studies presented in the book, it certainly helps if salaries have a variable component that is related to each collaborator meeting the individual, departmental and corporate objectives.

Showing results early is very important, giving the process additional external credibility and boosting employee morale. For instance, in one of the case studies in the book, water loss was halved in just eight months, an achievement that received great coverage in the media.

In every process of change there will inevitably be large and small obstacles to overcome. It is crucial to evaluate them carefully, to gather sufficient strength to surpass them and to wait for the right moment to take action. The most frequent and difficult obstacles are related to people. The prospects opened by the change management process itself are likely to induce self motivation in the best collaborators, but human nature can make it difficult for some of them – for instance those directly responsible for the previous poor performance– to be really committed to the new project.

The change management process requires that the previous organizational structure be broken down and that many procedures and routines are abandoned. Leadership is essential to make sure that the collaborators keep their motivation and help each other achieve the new corporate objectives. Specific restructuring projects with suitable team leaders may replace much of the former organization, but there is a time limit for turbulence without a stable organizational chart.

At some stage, when major objectives are encountered, it is necessary to stabilize a new organizational structure and to consolidate what has been achieved. It is the time to anchor the new behaviour in the new organizational structure. It is important to stress that managing continuity and change management towards excellence are two completely different processes, requiring different approaches, methodologies, organizational structures and project leaders. Excellent companies, especially if they achieved excellence some considerable time ago, should certainly be studied for best practices, but they may not be the most effective to provide guidance to poorly managed ones aiming to improve. Indeed, in such companies, activities are already segregated and stabilised, procedures are well established and documented for virtually every task; an organizational culture has been consolidated and cannot easily be transferred elsewhere.

From personal experience, it is important that the leader of a change process adheres to a clear set of rules, for instance those noted by John Kotter with the following hints:

(1) Adopt a simple and flat organizational structure, defined initially only up to the first level and later on, as required, to the second level of coordination;
(2) Set a restructuring team to coordinate the change process and meet them regularly as a group;
(3) Define restructuring projects for the key objectives and assign each one to a member of the restructuring team;
(4) Trust the team members but avoid dispersion: give them freedom, but not the authority to alter the vision, the objectives, the restructuring projects or the priorities;
(5) Plan and act in short intervals and avoid lengthy studies (a project leader with previous experience and who knows the water utilities field needs fewer studies to make decisions);
(6) It is natural and therefore necessary to accept a certain amount of chaos during change management processes due to the fragmentation of the former organization.

Change management is by definition a transition, and the more intense phase of this change cannot take more than a few years. However, after this initial period, change really means improving and it must be go on, although necessarily at a slower pace. In his book "Good to Great", Jim Collins (Collins, 2001) picks up on the common underlying forces behind successful sustainable processes of change in international companies. Amongst the 1,500 companies listed in Fortune magazine between 1965 and 1995, Collins identified only eleven companies with sustainable growth for at least fifteen years. Collins classifies leaders in five levels. Level 5 refers to the top of a hierarchy of executive capabilities. Leaders at the other four levels may be successful, but may not be able to lead companies to sustained excellence. According to Collins, every one of the good-to-great companies had level 5 leaders in the critical transition phase.

Collins recommends that the next step in the change process should not be the definition of a new vision for a company, as is often suggested, but rather the selection of the team, taking on board the right people and putting them in the right places. It is with this team that the leader should face the reality encountered by the utility and define the strategy. It is important to combine discipline with creativity within the team: excessive control is not necessary if the action is disciplined and if disciplined thought helps control creativity excesses that do not contribute to the corporate objectives.

Finally, it is important to realize that technology is a powerful accelerator but that it is never the main instrument of the transformation process of a utility. The best water companies are not necessarily pioneers in the application of new technologies, but they are certainly eager to adopt well proven ones, even taking into consideration the corresponding time and learning costs, which are usually very high.

#### *2.2.2.2 International change management experiences*

Regulators ands international water organizations systematically present the results of benchmarking studies of water utilities. Increasingly, they tend to use IWA or other similar performance indicators, and this fact on its own is extremely important to allow for meaningful comparisons between utilities and between studies. In a World Bank publication (Ginneken Kingdom, 2008), based on case studies on well performing utilities, it is concluded that there is no perfect model for public utilities that guarantees good performance, but well functioning utilities share the following main attributes:

- Autonomy, accountability and customer orientation; and
- Separation of the political realm from the service provision (although full isolation is neither attainable nor desirable).

The study also refers to the following attributes which essential for successful change processes:

- The appointment of a champion within the utility who functions as an agent of change;
- Reforms should be incremental and focus on areas where prospects for early success are high, with tangible results within one political cycle; and
- Best-fit is more important than best-practice (each case is unique).

#### *2.2.2.3 The first 100 days of the change management process*

The first 100 days in a change management process are critical. There is not a second opportunity to create a first good impression. Every word, every gesture, every behaviour defines the leadership's future. This period of time is both a window of opportunity and an increased risk. The opportunity results from the fact that decisions taken during these first three months will be less likely to be contested, internally or externally.

Thomas Neff and Jim Citrin, in their book "You're in Charge, Now What?" (Neff Citrin, 2007), expose an action plan for the first 100 days in a leadership position, with eight steps that include the preparation of the leader, alignment of expectations of stakeholders, formation of the management team, communication inside and outside of the organization and, most importantly, avoiding like disrespecting the predecessor, misreading the organization or succumbing to the saviour syndrome. Ideally, the mandate should be regarded as a series of successive first 100 days, each one with its own plan and goals, to keep the flame alive. In any case, the results obtained in the first three months are crucial to give external credibility and internal motivation to carry out the change management process.

### 2.2.3 Stages of development of water companies

#### *2.2.3.1 A rating scale with four levels*

Water utilities are primarily expected to provide good or excellent services. Quality of service may vary widely from company to company and makes the whole difference: for instance, providing a continuous high quality service, 365 days a year, 24 hours a day, requires an organization and tariffs that are totally different from those in a situation where interruptions and failures are still tolerated by customers.

Although the development of a water company is a continuous process, a set of qualitative indicators determines its positioning on a performance rating scale with four stages, as follows: pre-efficacy, efficacy, efficiency and, finally, excellence (Figure 2.4). When a water company starts to be professionally managed, performance indicators tend to show impressive improvements. The speed with which each company evolves along this scale, in pursuit of excellence, depends on the quality of its management and its specific context.

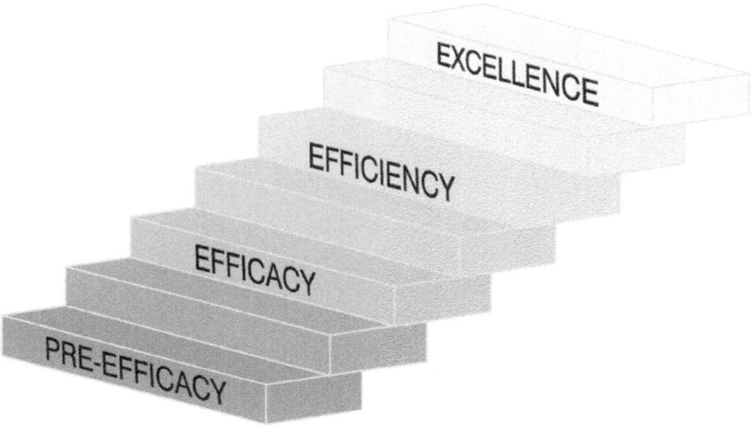


**Figure 2.4** Stages of development of a water company.

Pre-efficacy is essentially a euphemism to describe a water company that is not effective in doing what it is supposed to do; for instance, featuring a lack of network coverage, frequent supply interruptions, poor water quality or excessive customer complaints. Efficacy means that the company fulfils its major objectives but using too many resources. As a result, tariffs are higher than they should be or the company requires public transfers (lost fund subsidies) to survive. Efficiency is a lengthy stage of improvement and indeed is a never-ending process that leads to resource savings in every area of the water utility. Excellence is the ultimate goal, the unachievable perfection every manager should aim to reach. Reliability, innovation, special attention to customers, social responsibility, employee welfare, certification and sustainability are certainly amongst management concerns at an excellent water utility, where the basic problems are already under control.

The evolution of a water utility is a continuum and there is some degree of arbitrariness in any classification, but it is possible to use simple indicators to position any water company in this rating scale. Take, for instance, a company that does nothing specifically to control NRW, that only calculates it at the end of the year to include the figure in the annual report or, even worse, just estimates it, typically loses more than half of the water that enters the system. NRW is a very convenient indicator to classify water companies: it is widely used, easy to understand and says much about an organization, as it is necessary to mobilize the whole utility to improve it (Farley, 2001; Brothers, 2003; Farley Trow, 2003; Lambert, 2003; Waldrom, 2005; Alegre Almeida, 2009). Much thought has been put into the selection of the best NRW indicators to use for this specific purpose. It was decided to use both the percentage of the water entering the system and the ILI – Infrastructure Leakage Index. The former has many disadvantages and it may be misleading under certain circumstances, but it is easily understood by lay people and the media and it takes into consideration all types of losses, including apparent losses. There are extreme examples of excellent water utilities with 80% NRW and many others report economic levels of loss above 30%, but, in general, any value above 50% should be considered very high.

On the other hand, the ILI is valid only for real losses but is technically much better supported. For instance a water utility with an ILI of 2 has real losses which are two times higher than the minimum value that it would be possible to achieve through acknowledged best practices in well managed water utilities. It is recommended that the reference ILI values are doubled in developing countries in relation to developed ones (World Bank Institute, 2005; Liemberger McKenzie, 2005). Decision times, the availability of a central control room, time intervals at which system inputs and outputs are evaluated and the existence of active leakage control teams help to position a water company in the aforementioned classification (Figure 2.5 and Table 2.1).

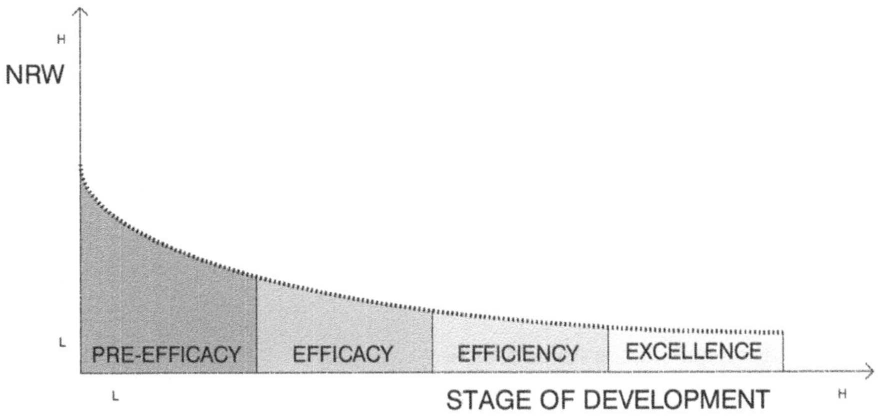


**Figure 2.5** Stages of development of water utilities and NRW.

**Table 2.1** Stages of development of water utilities

| Stages of Development Indicators | Pre-efficacy | Efficacy | Efficiency | Excellence |
|---|---|---|---|---|
| Decision times | Months | Weeks | Days | Hours |
| Calculation of Key Performance Indicators | Yearly | A few times a year | Monthly | Daily |
| Control Room | No | Office hours | 24 Hours a day | 24 Hours a day |
| Control of Water Inputs and Outputs | No | Monthly | Daily | Real time |
| Active Leakage and Pressure Control | No | Occasional | Systematic | Systematic |
| Control of Night Flows | No | No | Daily | Real time |
| Response to Clients and Bursts | Month | Week | Day | Hour |
| Non Revenue Water* | Very high (> 50%) | High (30 – 50%) | Moderate (15 – 30%) | Low (<15%) |
| ILI** | (>8) | (4-8) | (2-4) | (<2) |

** These values of NRW in percentages are only indicative and may vary widely according to the local conditions, namely pressure, production cost of treated water and age of the network. **The ILI reference values should be doubled in water utilities in developing countries, according to Word Bank recommendations (World Bank Institute, 2005).*

Figure 2.6 shows the main measures required to proceed along the successive stages of development of water utilities. In real water utilities, there are always measures in operation corresponding to any of the stages, but it is more likely that the basic ones have been taken at an earlier stage and that the most expensive and least cost-effective have been taken later.

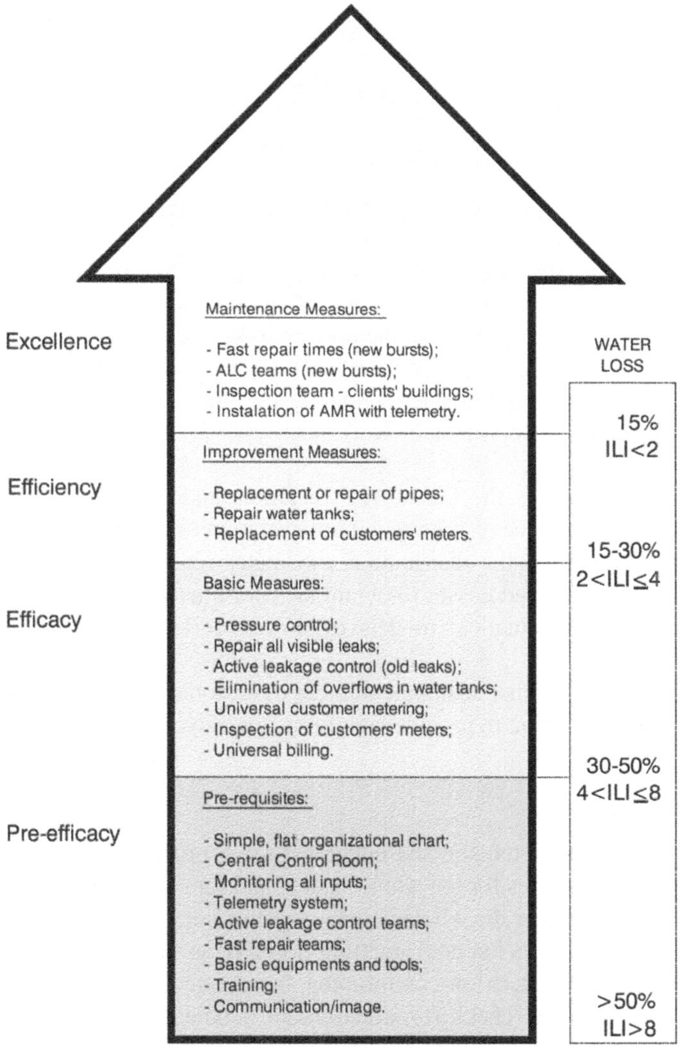


**Figure 2.6** Measures to reduce NRW towards ELL and sustainability.

### *2.2.3.2 Measures for efficacy, efficiency and excellence*

There are no absolute rules to manage a water utility in search of excellence. Each company has a history and no experience is directly applicable without the necessary adjustments: best fit is better than best practice, as previously mentioned. However, from personal experience, there are a number of measures that may be useful for leaders of change management processes, depending on the stage of development of each utility:

## Efficacy

1. Read everything about the company and other companies that may be related to it, namely news, previous audits and regulator reports;
2. Consolidate and motivate the teams responsible for ensuring, on the ground, the continuity of water services, primarily those teams which deal with emergency cases;
3. Identify expectations and possible reasons for dissatisfaction of shareholders, customers and employees;
4. Carefully select the key collaborators, directors and coordinators;
5. Create easy and fast communication routes with all collaborators;
6. Acquire all necessary safety and operational equipment, providing training in their use to prevent accidents and improve the working conditions of employees;
7. Identify and correct situations of wage inequity as soon as possible;
8. Create a variable component of the salary based on merit and results;
9. Create a control room that is available 24 hours a day, 365 days a year;
10. Replace manual and uncoordinated worksheet-type procedures with centralized auditable computer applications;
11. Install a GIS that includes the customers' locations and all available information about the networks, prioritizing service reservoirs, pumping stations and principal water mains;
12. Connect each customer to the distribution and meter reading zones and to the infrastructure database using their postal addresses to obtain the coordinates of buildings;
13. Develop simplified mathematical models of the networks using open source or proprietary software;
14. Evaluate the possibility of increasing the number of customers using existing infrastructure;
15. Prioritize network expansion to provide conditions for universal access to water services.

## Efficiency

16. Implement a long-term, systematic and integrated set of procedures for NWR reduction, starting with standard water balances for the whole system and for each of the distribution zones;
17. Develop an energy audit for the whole company, prioritizing major equipment, namely pumping stations and treatment plants but not forgetting (to serve as an example of efficient use of energy) the buildings and headquarters (air conditioning, lamps, etc.);
18. Optimize the labour force: check for situations of deficient organization and overstaffing and, where that is the case, attempt to replace external contracts with internal employees. Note that this is a very complex task that requires negotiation and training;
19. Develop comprehensive mathematical models of the distribution networks, linked to the corporate GIS;
20. Create conditions to bill and collect all services, to improve purchasing, and create internally or contract out the services of a project management structure, to contract and improve supervision of construction works, keeping schedules and budgets under control.

Excellence

21. Increase system reliability, aiming for 100% continuity of water services even in the event of bursts and emergencies;
22. Develop all aspects related to service quality, mostly covered by internationally accepted performance indicators;
23. Pay special attention to customers, including introducing a CRM – Customer Relationship Management-software, pro-actively avoiding reasons for complaints and ensuring these are resolved fast;
24. Encourage and reward changes of attitude of employees towards teamwork, cooperation and customer attention through motivation and training;
25. Move towards paperless procedures throughout the company;
26. Integrate computer applications as much as possible and extend GIS use throughout the company.

## 2.3 ECONOMIC SUSTAINABILITY OF WATER UTILITIES

### 2.3.1 The TTT sustainability model

As previously mentioned, water tariffs should be sufficient to pay for all the investment and operational costs in economically sustainable water utilities. However, in most countries, at least at some stage, it has been necessary to transfer funds to water utilities to support infrastructure construction. This was certainly the case in the European Union, where member countries had to accelerate their investment plans in order to meet increasingly demanding European environmental directives. In some regions, for instance with very small or dispersed populations, in mountainous areas or on small islands, the operating costs per capita may be too high to be fully covered by socially acceptable tariffs. Under these conditions, it may be necessary to call for regional or national solidarity to subsidize the operation of the water companies.

Economic sustainability of water utilities is, therefore, the result of a financing package, known as TTT (Tariffs, Transfers and Taxes). Figure 2.7 shows the financing model for sustainability, including the flux of money between water companies, regional and local authorities, government, international financial institutions, banks and the private sector (OECD, 2009). The international financing institutions, banks and private sector all participate in the process mainly through loans, guarantees, bonds and shares. However, this participation, although invaluable to accelerate investment programs, implies repayment, with interest or some other legitimate benefit, hence is not part of the basic source funding of the TTT sustainability model.

In many cases, water companies cannot obtain sufficient funds to finance their activity and tend to postpone investments mainly on renewal of systems, which is more necessary as they get older and closer to the end of their service life. This infrastructure deterioration is a very serious problem that impoverishes a water utility, reduces its reliability and leads to serious future problems. On the other hand, borrowing and Public Private Partnerships (PPP) postpone the payment of the debt service to the future generations, corresponding to investments that were made earlier. Indeed, in both cases, future generations, without being consulted of course, act as a fourth basic component of source funding in the above mentioned TTT financing model, in addition to the tariffs, transfers and taxes. Although they will also benefit from them, they may end up paying more than their fair share of investments in the infrastructure they have inherited.

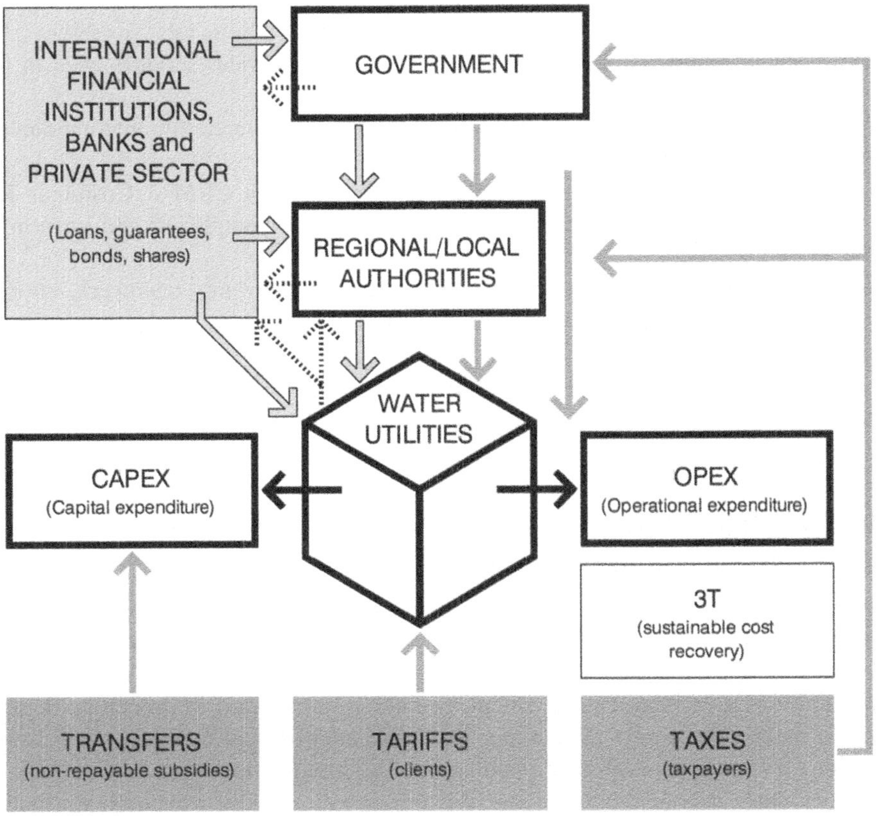


**Figure 2.7** The TTT sustainability model of water utilities (Source: OECD, 2009).

## 2.3.2 Macro and micro sustainability

There are two different types of sustainability problems: macro sustainability and micro sustainability ones. Many water utilities around the world do not, or did not in the past, fully recover both investment and operation costs, leading to what is called a macro-sustainability or macro-affordability problem. Alternatively, in any water utility there are always customers who cannot afford to pay their water bills, leading to a micro-sustainability or micro-affordability problem.

In some cases, particularly in developed countries, there are solid grounds for setting tariffs at a sustainable level. However, in developing countries, the scope for full cost recovery is severely limited by widespread poverty. To be affordable, it is often considered that tariffs should not represent more than 3% of the average household income. However, this affordability yardstick does not take into consideration the different income levels of the customers. For instance, tariffs may have a weight of less than 3% in the average family budget, but for the poorest families it may represent much more.

Unfortunately, experience shows that it is very difficult to design, and even harder to implement, a system to subsidise poor families selectively and often these are the ones that end up paying more for water. Keeping tariffs low for everybody does not necessarily help the poor and especially those who currently do not have access to water services. Unsustainable low tariffs, in fact, may prevent the extension of the network to poor communities, forcing them to continue paying higher prices for water of inferior quality.

The solutions to affordability fall essentially into two main categories: subsidizing poor families or subsidizing water bills that the poor families pay. The first solution operates through the social security system and is external to the water company. The second is based on tariff-related measures that try to keep poor families' water bills low through some form of cross-subsidization and may be designed and implemented within the water company itself. It is not easy, however, to subsidize consumption of poor families selectively. For instance, the common practice of increasing block tariffs, for example monthly supplying the first five to ten cubic meters at low cost, in fact favours people living alone (often not poor) and the owners of weekend or holiday houses (certainly not poor) and penalizes large families with high consumption (who may or may not be poor).

In addition to cross-subsiding within the domestic sector (large consumers paying more than small consumers), there is also often cross-subsidization between the non-domestic sector and the domestic sector, with industrial and commercial users frequently paying higher tariffs. In water utilities, there are often complex cross subsidizing tariff structures: the decision on how much each category of consumers should pay is, in these cases, of a political nature. However, when there is a regulator, they tend to interfere in the process of tariff setting, eliminating cross subsidizing as much as possible and pressing towards more similar rates to all the customers.

### 2.3.3 Financing of water services (always insufficient)

To provide water supply and wastewater services, water utilities need very high financial resources to cover their investment and operation costs. However, this sector has been severely under-funded in many places for many years, with serious financing gaps, especially in developing countries, though in the developed countries there are also great financing requirements to replace ageing networks and treatment plants (Figure 2.8).

**Figure 2.8** Required investment in infrastructure, 2005–2030 (trillions of $US; source OECD, 2009).

Current spending is significantly below the required levels and the consequences are clearly visible:

- Under-financing of water utilities;
- Inadequate operation and maintenance of infrastructure;
- Cuts or delays in the implementation of investments;
- Deterioration of water and wastewater systems;
- Loss of service quality; and
- More interruptions of supply and failures.

The huge investment gap, together with poor management, explain that in spite of the ambitious objectives set in high level international conferences on water and sustainable development since the 1977 UN Conference on Water for the Nations in Mar del Plata, access to potable water and wastewater services is still not a reality for more than one billion people in the world. The last global objective, included in the United Nations Millennium Declaration of 2000, was set to halve the proportion of people without access to water and sanitation by 2015 (UN, 2008; World Bank, 2008). This target is less ambitious than previous ones but, unfortunately, it is certain that it will not be met in many parts of the world.

These global meetings and conferences on water, attended by heads of state and tens of thousands of water stakeholders from all over the world are very important to put water on the political agenda, to grab the attention of the general public for water as a human right and as a global problem, to define the magnitude of the resources that are necessary and to set future objectives; however, they seldom lead to an effective allocation of funds for international assistance. These global objectives must always be regarded as indicative and, to be achieved, they should be in line with the availability of funds and with customers' ability and willingness to pay, leading to economically sustainable solutions. In some cases, this is at least approximately possible, in others not so much, mostly because of poverty, poor governance and poor management.

Poverty-related problems, of which access to potable water is certainly a major one, are greater in the slums of the megacities of the developing countries, but they occur indeed everywhere in the world. In fact, in every water utility there are always people who cannot pay their water bills, for limited periods of time or even for years. It is certainly important to find global solutions as much as possible, but local solutions are always required to solve peoples' problems, including the new forms of poverty that emerge in times of a financial crisis. Global solutions are always extremely difficult to achieve because there are great asymmetries in the distribution of world resources per capita, not just between countries, but within the same country, whilst there are countries and regions with huge gaps in terms of water infrastructure.

The so-called improved water services, namely systems which improve on the existing situation, whilst still missing some components of a conventional urban system, although not ideal, are of course an affordable improvement on the existing ones. On the other hand, experience shows that implementing excellent but non-sustainable solutions is a waste of money and in the end does not solve people's problems.

The following simple common sense methodology may be used to create sustainable water utilities in situations where it may not be possible, initially, to implement water services that meet all the developed country standards:

(1) Define the maximum amount of transfers and taxes that might realistically be expected (international, national or regional solidarity);
(2) Define current and future affordable tariffs (local study);
(3) Define investment required to provide an improved level of service (local study);
(4) Design organization to undertake rehabilitation and operation of the system (local study);
(5) Determine the minimum tariffs required;

(6) If the minimum tariffs required exceed the affordable tariffs, it is not possible to implement a sustainable solution; in that case, attempts should be made to try and attract additional funds or accept less ambitious objectives;

(7) If the tariffs required are lower than the affordable tariffs, it is possible to go for more ambitious objectives.

## 2.4 SEARCHING FOR SUSTAINABILITY: THE PORTUGUESE DUAL SYSTEM

In 2014, Portugal, with a population of around 10 million people, has a modern water supply and wastewater system that covers almost all of the country. It was installed mostly in the 20 years from 1993 onwards and, as has happened in other EUmember states, it has benefitted from important transfers of European funds. The evolution of the water quality in the water supply systems all over the country has been impressive and approaches 100% (Figure 2.9).

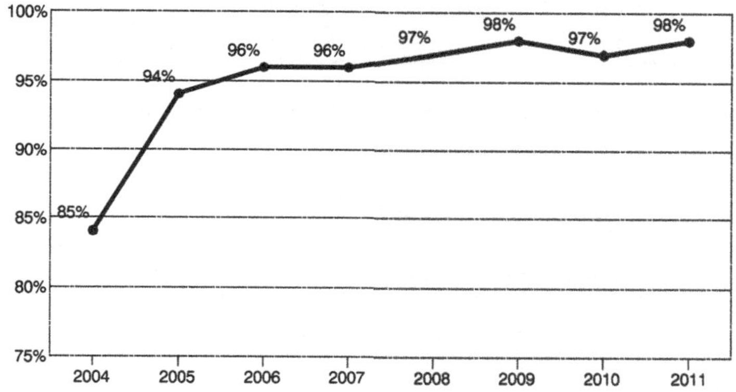


**Figure 2.9** Evolution of water quality in public water supply systems in Portugal (Source: ERSAR, 2013).

The number of water sources was drastically reduced from about 12,000 to 2,000, including a small number of modern and very reliable water treatment plants that serve hundreds of thousands or millions of people. The territorial coverage of public water supply and wastewater services also increased very fast and is already almost universal, except for some dispersed and isolated populations, served by individual systems (Figure 2.10).

In the early years of EEC membership, the rule was still that each municipality would plan, design, build and operate its own water supply and wastewater system. The size of municipalities and consequently of the water utilities presented a wide variation, and the majority of them are very small (Table 2.2). In 1975, the water sector was nationalized and private companies were not allowed to manage water supply and wastewater systems. In 1993, this situation changed in two main aspects: municipalities could again grant the concession of their water and wastewater systems to private companies, whilst a dual system started to be implemented, with the creation of regional multi-municipal bulk water supply and wastewater treatment companies. (Pocas-Martins, 2007). A public national holding company, ADP - *Águas de Portugal* - was created, owning the majority of the shares of each of the multi-municipal companies; the other shareholders of each multi-municipal company were the municipalities served

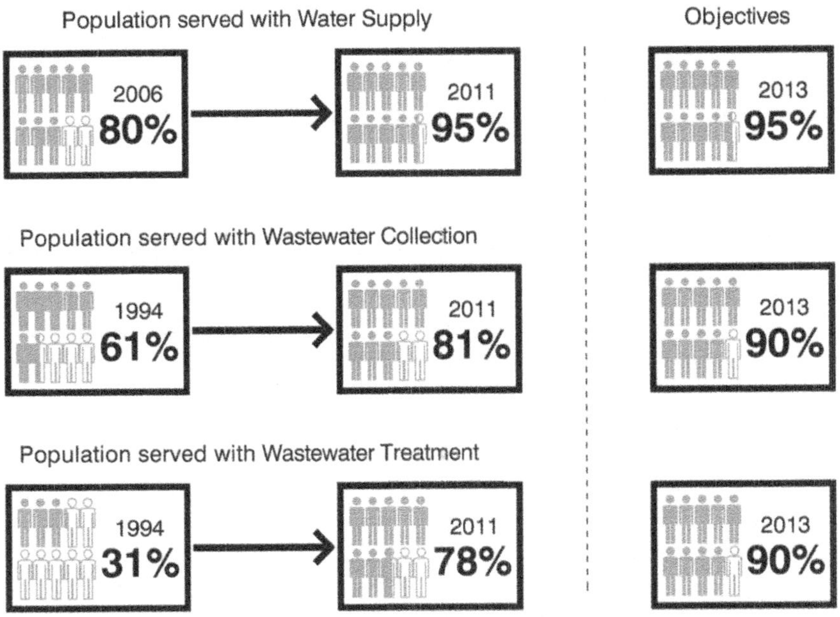


**Figure 2.10** Evolution of the population served with water services in Portugal (Source: ERSAR, 2013).

**Table 2.2** Size of water utilities in Portugal (Source ERSAR, 2013).

| No. Customers | No. Utilities | % Utilities | Resident Population | Domestic Customers |
|---|---|---|---|---|
| <5,000 | 91 | 32.0% | 5.2% | 5.9% |
| 5,000 to 20,000 | 127 | 44.7% | 24.0% | 23.0% |
| 20,000 to 50000 | 45 | 15.8% | 27.0% | 26.3% |
| 50,000 to 100,000 | 12 | 4.2% | 16.6% | 15.8% |
| >100,000 | 9 | 3.2% | 27.3% | 29.1% |

by each system. Management of the multi-municipal bulk water and wastewater systems was therefore controlled by ADP, and these systems were operated directly with different levels of outsourcing.

Water distribution and wastewater collection continued to be under the responsibility of the municipalities, which could operate them directly, grant their concession to private firms or, from 1998, create municipal companies, controlled by the public shareholder. The majority of these municipal water companies were totally owned by the municipalities, but some had a minority private shareholder, typically responsible for management. In 2013 around 28% of the Portuguese population had their water supply and/or municipal wastewater services managed by private operators. The evolution over time, from 1993, is presented in Figure 2.11.

In 2014, most of the country is covered by regional bulk water companies, according to Figure 2.12, which also shows that the vast majority of the municipalities have opted to operate their own water supply distribution and wastewater collection systems directly.

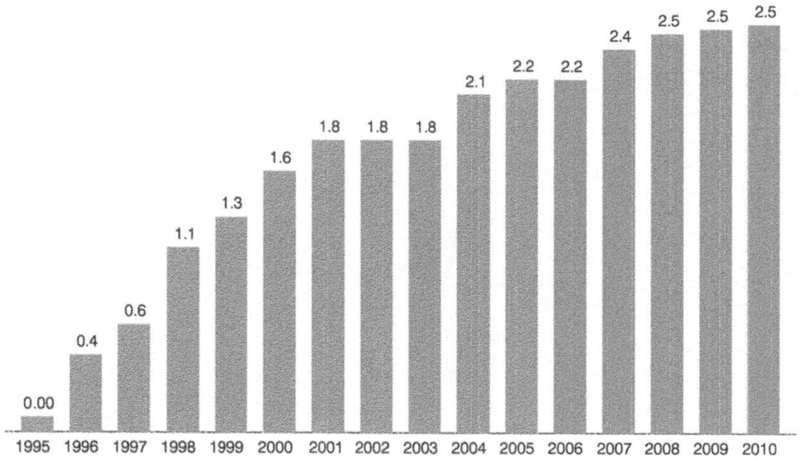


**Figure 2.11** Evolution of the Portuguese population served by private operators, in millions (Source ERSAR, 2013).

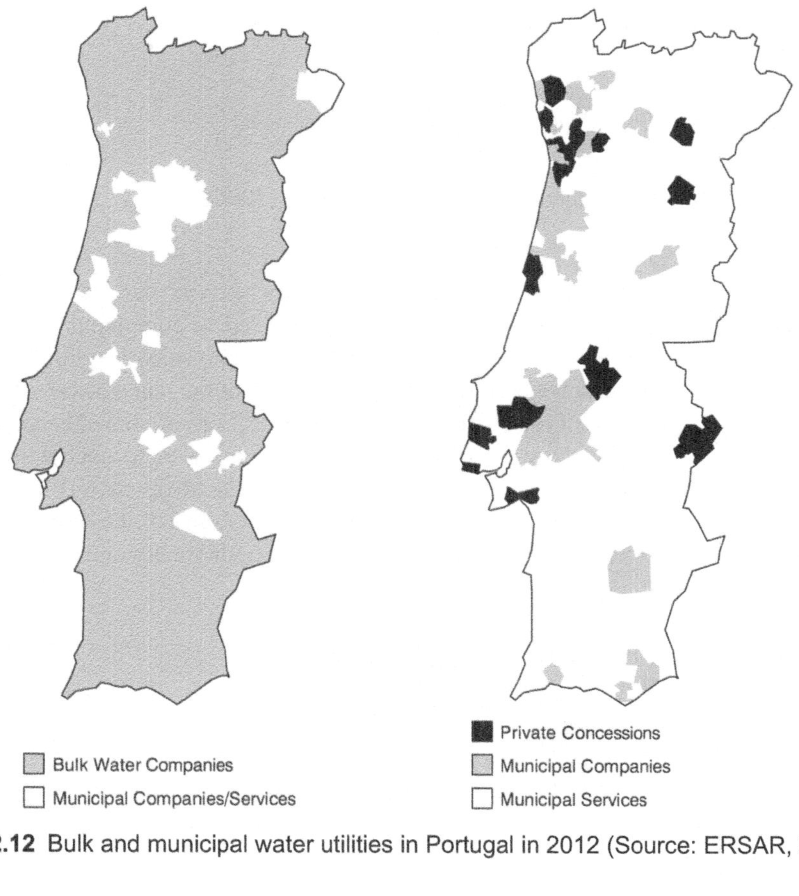


**Figure 2.12** Bulk and municipal water utilities in Portugal in 2012 (Source: ERSAR, 2013).

The first bulk regional water companies, created in the early 1990s (Figure 2.13), served the most populated urban areas of the country; their success led to a political decision to expand the same model to rural areas, where the population was lower and often scattered.

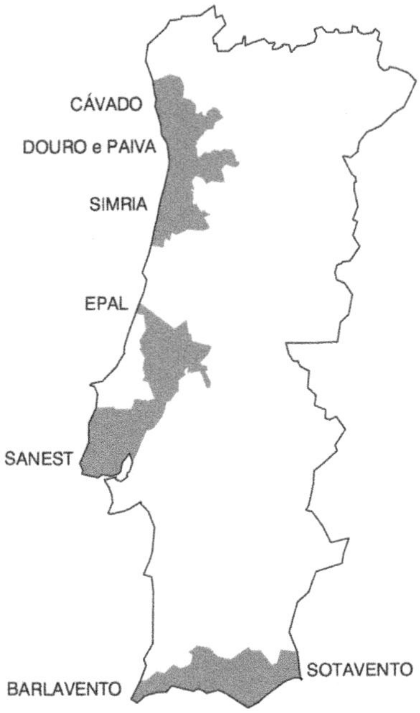


**Figure 2.13** The first generation of bulk water utilities in Portugal (1996).

As would be expected, tariffs in the bulk water companies serving the inland areas resulted in significantly higher charges than those in the major cities. This led to complaints and debts from the municipalities. As a result, the financial situation of some bulk water companies deteriorated. Unfortunately, the financial difficulties of these bulk water utilities are symptoms, not the real disease. In fact, there is an underlying problem: the high value of the tariffs that would be necessary to make the smaller municipal water utilities sustainable under the current conditions. This value varies very much with the dimension of the utilities and may reach about €10/m$^3$, five times more than in the urban areas (Figure 2.14). In fact, in regions of low population and disperse housing, investment costs to install the water supply pipes and sewers are much higher whilst, on the other hand, tariffs have been traditionally much lower (Table 2.3).

**2-10 €/m$^3$**

Efficiency and Solidarity

**Figure 2.14** Average municipal water and wastewater tariffs required for the sustainability of water utilities.

**Table 2.3** Relative variation of water and sanitation bills for domestic Portuguese customers (in relation to the consumption of $10m^3$/month in a large city).

| Population (thousands) | Water and sanitation bill for domestic consumers (Euros/month) | | | |
|---|---|---|---|---|
| | $5m^3$/month | $10m^3$/month | $15m^3$/month | $20m^3$/month |
| > 100 | 0.5 | 1.0 | 1.4 | 2.2 |
| 50 – 100 | 0.4 | 0.8 | 1.2 | 1.8 |
| 25 – 50 | 0.4 | 0.7 | 1.0 | 1.4 |
| 10 – 25 | 0.3 | 0.6 | 0.9 | 1.3 |

There are also known economies of scale in the water supply and wastewater sector: it is much more expensive to operate a system that serves a small population than one that serves a big city. All across the country, in 1994, at the beginning of a major investment phase in water infrastructure, there was an inverse correlation between investment per capita required to install water infrastructure and the tariffs charged to customers. Each of the municipalities is represented in Figure 2.15 by a circle proportional to its population.

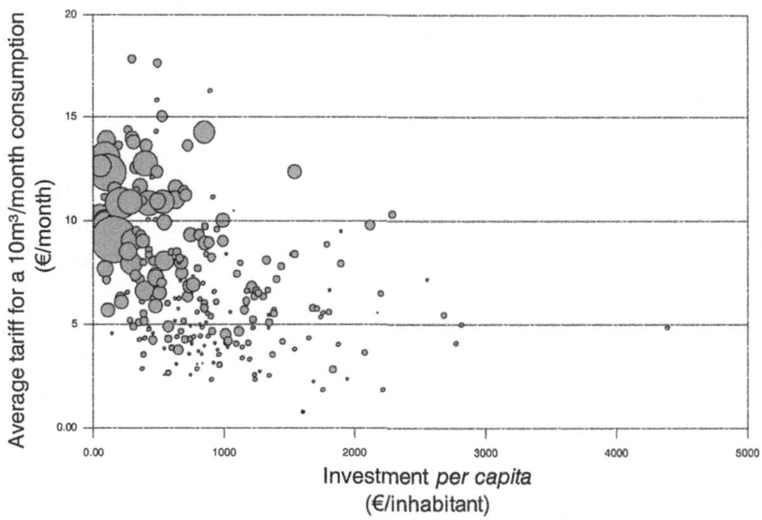


**Figure 2.15** Correlation between water and wastewater tariffs and investment *per capita* required in Portuguese municipalities for an average family consumption of $10m^3$/month (Source: Pocas-Martins, 1998).

Due to the long-term tradition of subsidizing water and wastewater services, the lowest tariffs are encountered in the smaller inland municipalities, where investments in infrastructure *per capita* and operating costs are higher. This reflects an asymmetrical distribution of the population and an excessive number of small water utilities, leading to a complex sustainability problem. Aggregating municipal services to create regional water and wastewater utilities is an obvious possible solution, but this is politically very difficult to impose and time consuming to negotiate. Maintaining the water supply distribution and wastewater collection, including billing to the final customers, within municipalities and creating bulk regional water utilities for abstraction and water treatment and/or for wastewater treatment and disposal was considered as an intermediate solution that was politically easier to accept and faster to implement. The aggregation of municipal services is easier to negotiate when there are national or international lost fund subsidies/transfers available. The opportunity of benefitting from these subsidies certainly helps to generate political consensus and win-win solutions.

From personal experience, there are no great problems in splitting water supply services, with a bulk water supply company selling potable water to the municipalities which, in turn, distribute it and charge customers. There are, of course, limitations to this compromise solution, namely exposure to municipal debts and the fact that it may not be possible to attempt a global optimal operation of the whole system. In fact, the optimal operation of the bulk water supply systems and of each of the distribution systems does not coincide with the optimal operation of the whole system. This dual solution should lead to lower levels of water loss in the distribution systems because in this case leakage is necessarily valued at the price the municipality acquires treated water, typically higher or even much higher than the corresponding marginal production cost. For instance in an integrated water utility that has plenty of raw water and spare treatment and carrying capacity, the net cost of producing an additional cubic meter of potable water corresponds essentially to only the energy and chemicals required, which leaves very little margin to invest in leakage reduction. In terms of wastewater systems, however, there are important disadvantages in splitting collection from the treatment and disposal components. It is necessary to install an expensive and not very effective metering system for billing purposes and it is difficult to manage infiltration flows and overflows, which may cause pollution in the receiving waters, namely urban streams, rivers and beaches.

In summary, this dual system may be an interesting option if it is necessary to implement regional water supply or wastewater treatment solutions in a region with many small municipalities in a short period of time. This may also be an interesting solution when local water sources are not available and it is necessary to find a regional water source outside the boundaries of those municipalities. However, the dual system does not necessarily, on its own, lead to economically sustainable systems and experience shows that it is important to address the sustainability issue from the beginning, before actually building the infrastructure. Forming systems that include large and small municipalities is certainly more likely to lead to sustainable solutions. If the small rural municipalities are left behind, it may be very difficult to mobilize regional, national or international solidarity later on, in order to bridge the gap between the affordable tariffs and the ones that are necessary to ensure the sustainability of such small utilities.

## 2.5 NRW AS PART OF THE SUSTAINABILITY SOLUTION

### 2.5.1 The IWA Water Balance

In the past, there were a number of different technical definitions of water loss, making it difficult to establish meaningful comparisons between water utilities. Fortunately, the IWA Water Loss Task Force developed a water balance methodology that has become internationally accepted, as described in Figure 2.16.

The designations of the major components of the IWA water balance are almost self-explanatory, as follows:

- System Input Volume (SIV) is the annual volume of water entering the distribution system (it may be defined in a different way when there is exported and imported water);

- Revenue Water (RW) is the annual volume of metered and non-metered water billed to registered customers;
- Non-Revenue Water (NRW) is the difference between System Input Volume and Revenue Water;
- Authorized Consumption is the sum of annual volume of water taken by authorized customers, either billed or unbilled;
- Water Loss is the difference between System Input Volume and Authorized Consumption and consists of Apparent Losses and Real Losses;
- Real Losses are the annual volume of water lost through all types of leaks, bursts and overflows on mains, service tanks and service connections, up to the point of customer metering;
- Apparent Losses consists of unauthorized consumption and all types of metering inaccuracies.

Ideally the System Input Volume should be equal to the Billed Authorized Consumption, and all the other components of the right-hand column of the water balance should be approximately zero. However, in reality, this never happens. To start with, it is virtually impossible to meter every single customer and estimates have to be made to account for the lack of meters or their respective reading for authorized customers. Therefore, the Billed Unmetered Consumption is always greater than zero and, if it is

| | | | | |
|---|---|---|---|---|
| SYSTEM INPUT VOLUME | AUTHORIZED CONSUMPTION | BILLED AUTHORIZED CONSUMPTION | Billed Metered Consumption | REVENUE WATER |
| | | | Billed Unmetered Consumption | |
| | | UNBILLED AUTHORIZED CONSUMPTION | Unbilled Metered Consumption | NON-REVENUE WATER |
| | | | Unbilled Unmetered Consumption | |
| | WATER LOSSES | APPARENT LOSSES (commercial) | Unauthorized Consumption | |
| | | | Customer Meter Inaccuracies & Data Handling Errors | |
| | | REAL LOSSES (physical) | Leakage on Transmission and Distribution Mains | |
| | | | Leakage and Overflows from the Utilities Storage Tanks | |
| | | | Leakage on Service Connections up to the point of Customer Meter | |

**Figure 2.16** The Water Balance table developed by the IWA Water Loss Task Force (Source: IWA Water Loss Task Force).

significant, it affects the precision and therefore the practical use of the water balance for developing a NRW reduction strategy.

The Unbilled Authorized Consumption refers to water that is supplied at zero cost to authorized customers, either metered (including social organizations and municipal services) or estimated (fire fighting). The water used by fire brigades is almost never measured for two reasons: it would be very expensive to install meters at each hydrant and the installation of such meters would reduce the flow during fire fighting when it is most needed. In some cases, for political reasons, water is offered to many customers: poor people, social organizations and even industries that municipalities want to attract to their territory. Also, in the slum areas of some cities, there is a generalized and tolerated (implicitly authorized) habit of stealing water from water mains. In normal conditions, in developed countries, this Authorized Billed Consumption is negligible but elsewhere there are situations in which this is a major component of the water balance.

NRW is the sum of Authorized Unbilled Consumption and Water Loss, which in turn is the sum of Apparent Losses (Unauthorized Consumption and Meter Reading Errors) and Real Losses (in Mains, Connections and Service Tanks). In fact, as previously mentioned, NRW can be regarded as a common symptom of seven "diseases", each with its own therapy: Billed Metered Consumption, Unbilled Unmetered Consumption, Unauthorized Consumption, Meter Reading Errors and Real Losses in Mains, Connections and Service Tanks. It is crucial to realize that combating NRW is essentially a management issue that requires the participation of the whole water company and involves the evaluation and reduction of each of its seven components.

### 2.5.2 Reducing Non-Revenue Water

Non-Revenue Water is a meaningful indicator of the global efficiency of a water company. The higher the NRW, the higher the potential savings, but also the greater the depth and complexity of the process of change required. The savings obtained from the reduction of NRW may be sufficient to cover not only the costs of a change management process but also to partially finance the expansion and rehabilitation of infrastructure without increasing tariffs or increasing them less than it would be necessary otherwise. The management of any water utility aims to reduce the amount of NRW, which can be accomplished reducing the SIV – System Input Volume – and/or increasing the RW – Revenue Water – (Figure 2.17).

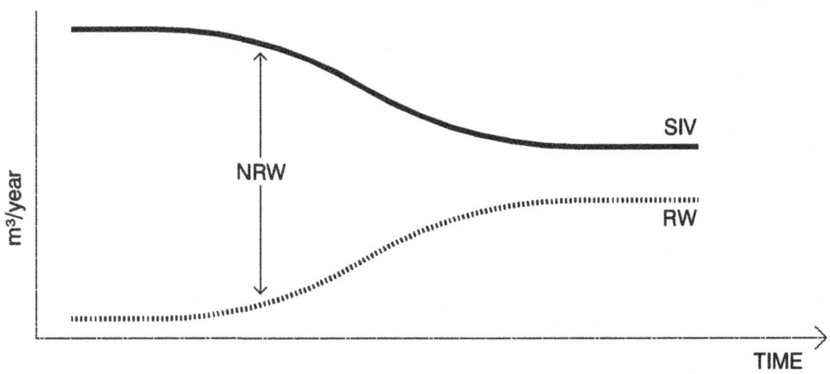


**Figure 2.17** Reduction of NRW over time. (SIV: System Input Volume; RW: Revenue Water).

Increasing one cubic meter of RW is worth a few times more than the reduction of one cubic meter of the SIV because the first is valued at selling price (including the wastewater service if this is related to water consumption, as is often the case), whilst the second is valued at acquisition or marginal production costs. The cumulative effect of reducing the SIV and increasing RW leads to a positive impact on the annual results of the water company (Figure 2.18).

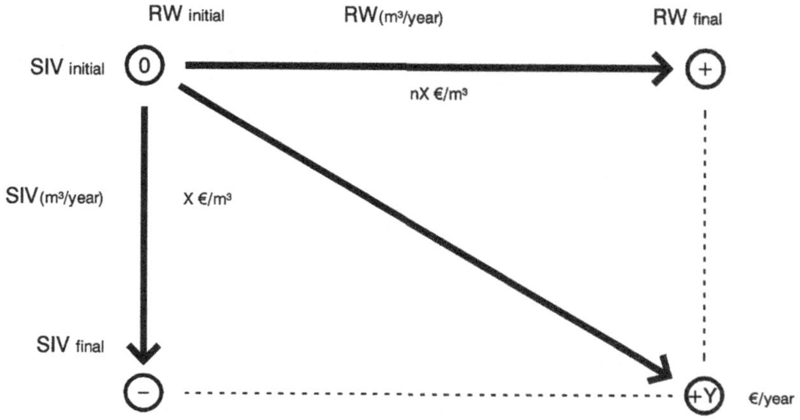


**Figure 2.18** Reduction of NRW: impact on the annual results.

The reduction of NRW can be accomplished through a number of activities that are well typified. Each of these activities is effective in reducing either real losses or apparent losses and has certain unit costs and benefits. In most cases, the cost of reducing one cubic meter of NRW may be assumed constant whilst the corresponding unit benefit typically follows a law of diminished returns. Both unit costs and benefits are specific to each water utility.

It is possible to develop a simple table that presents the seven components of NRW included in the IWA water balance and the activities that are effective to reduce each of them (Figure 2.19). Simple common sense dictates that any NRW reduction strategy should start with the less costly activities, for example active leakage control or pressure management, and leave the more expensive ones, like pipe replacement, to the end.

There are two basic questions when it comes to NRW reduction: how far should a water utility go in terms of NRW reduction and what should be done to get there. There is, of course, no single or simple answer to any of these questions. The first, related to target setting, is very complex because it depends on technical, economic, environmental, social and political options. The second question is essentially of a technical and economic nature and may also have different answers. In practice, investments to reduce NRW may be strongly determined by the regulator (if there is one) and by the owners' priorities (more or less pressure for short-term profit). Subject to these restrictions, water utilities may develop strategic plans to invest in the reduction of NRW, which are implemented through submission to the managing board of projects that include a set number of units of each of the above-mentioned activities. These projects are typically supported by cost-benefit analysis and refer to one or more distribution zones into which the system is divided. These cost-benefit analyses may be carried out over different periods and using different economic parameters, as will be detailed in the following chapter.

Reducing NRW should not be a goal in itself, and the concept of the Economic Level of Loss is still open to debate in many aspects, leading to different results. For example, one of the explanations for the average 5% rise of water losses in Chile after the transfer to private operators is that water utilities, under public management before privatization, may have reduced losses to below the optimal economic level (Ducci Medel, 2007 quoted from Marin, 2009).

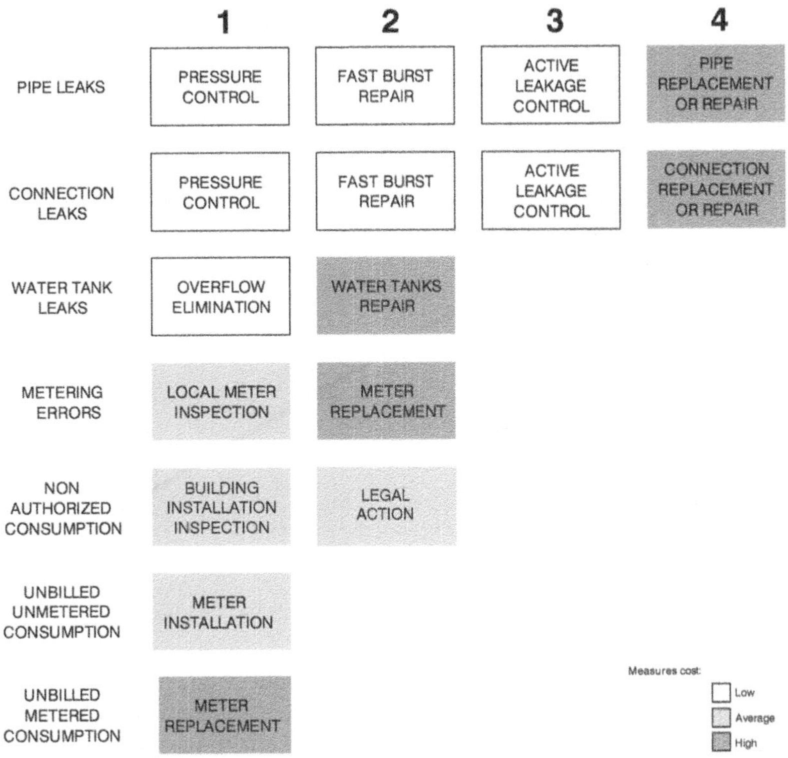


**Figure 2.19** Measures to reduce NRW.

Another interesting point concerning ELL is related to its wide variations from utility to utility. A European utility, for example, had an ELL above 80% of the System Input Volume because it received nearly potable water by gravity from pristine sources at very low cost (Donnelly A. (2014), *Personal Communication*).

In this book, a new approach is proposed for NRW target setting: the final value of NRW is not estimated a priori but, instead, may be calculated as a result of an optimization algorithm that minimizes the difference between the marginal benefits and marginal costs of implementing the activities that contribute to NRW reduction. Taking all relevant factors into consideration, both internal and external to the water utility, the top management of water utilities may use this methodology to decide on the appropriate target.

## 2.6 REFERENCES

Alegre H. and Almeida M. C. (2009). Strategic Asset Management of Urban Water Infrastructures. IWA Publishing, London, UK.

Brothers K. (2003). A practical approach to water loss reduction. *Water 21,* Magazine of the IWA, Issue 5.3, June 2003, pp. 54-55.

Collins W. (2001). Good to Great: Why Some Companies Make the Leap ... and Others Don't. HarperCollins Publisher, New York.

Ducci J. and Medel O. (2007). Chilean Experience with Private Sector Participation in Water Utilities, Mimeo, ETWWA, World Bank, Washington, D.C., September 2007.

ERSAR (2013). RASARP 2012, Lisbon, Portugal.

Farley M. (2001). Leakage Management and Control – A Best Practice Training Manual. World Health Organization, Geneva, Switzerland.

Farley M. and Trow S. (2003). Losses in Water Distribution Networks – A Practitioner's Guide to Assessment, Monitoring and Control. IWA Publishing, London.

Ginneken M. and Kingdom B. (2008). Key Topics in Public Water Utility Reform, World Bank, Water Working Notes, Note No.17, August 2008.

Jones C. (2010). Kotter's 8-Steps: Leading Change in the 21st Century Organization, http://sourcepov.com/2010/04/10/org-culture-intervention/, last accessed on 2014.05.30.

Kotter J. (1996). Leading Change. Harvard Business School Press, USA.

Lambert A. (2003). Assessing non revenue water and its components: a practical approach. *Water 21,* Magazine of the IWA, Issue 5.4, August 2003, pp. 50-51.

Liemberger R. and McKenzie (2005). Accuracy Limitations of the ILI: Is it an Appropriate Indicator for Developing Countries, Conference Proceedings, IWA Leakage2005 Conference in Halifax, Nova Scotia, Canada. http://waterloss2007.com/Leakage 2005/index.php, last accessed on 2014.05.30.

Marin P. (2009). Public–Private Partnerships for Urban Water Utilities: A Review of Experiences in Developing Countries. The World Bank and PPIAF Public–Private Infrastructure Advisory Facility, Washington, D.C., February, 2009.

Neff T. and Citrin J. (2007). You're in Charge, Now What? Crown Business Publisher, New York.

OECD (2009). Managing Water for All: An OECD Perspective on Pricing and Financing. Organisation for Economic Co-Operation and Development, Paris, France.

Pocas-Martins J. (1998). Public Water and Wastewater Services: Financing and Management Options in the Portuguese Municipalities (in Portuguese). AEPSA – Associação das Empresas Portuguesas para o Sector do Ambiente, Lisboa, Portugal.

Pocas-Martins J. (2007). The prices and sustainability test. In: WATER XXI – 14 Years of Change (in Portuguese), ed. APDA, Lisbon, Portugal, pp. 112-117.

Pocas-Martins, J. (2009a). Water Loss Reduction and Change Management, Paper presented to the IWA Water Loss 2009 Conference, Cape Town, South Africa.

Pocas-Martins J. (2009b). Water loss reduction and change management. *Water Practice and Technology*, Vol 4, No. 3, http://www.iwaponline.com/wpt/004/0051/0040051.pdf, last accessed on 2014.05.30.

United Nations (UN) (2008). The Millennium Development Goals, Report 2008, United Nations, New York, NY, USA.

Waldrom T. (2005). Management and Reduction Losses from Water Distribution Systems. Environmental Protection Agency and Wide Bay Water Corporation, Queensland.

World Bank and International Bank for Reconstruction and Development (2008). Global Monitoring Report 2008: MGDs and the Environment – Agenda for Inclusive Sustainable Development. International Bank for Reconstruction and Development and World Bank, Washington, D.C., USA.

World Bank Institute (2005). NRW Training Module 6: Performance Indicators, pro-wat.com/uoloads/media/prowatwlbook_english.pdf, last accessed on 2014.05.30.

# Chapter 3
# The Efficiency Challenge

## 3.1 INTRODUCTION

On the road to excellence, the longest stage is certainly that relating to efficiency. Indeed there are no limits for using less and less resources to do the same things or to use the same resources to do more things. Indeed, no matter how efficient a water utility is regarding a certain performance indicator, it can do better and improve it. However, the more efficient the water utility becomes, the harder it is to identify measures to further increase efficiency that make economic sense, that is, for which the improvements are more valuable than the corresponding additional inputs.

An efficient water utility will undoubtedly have a low NRW value, uses only the energy that is necessary and possesses an adequate number of employees whilst outsourcing is also kept to a minimum, consistent with the provision of a pre-defined high level of service quality. Its efficiency may be measured through a number of standard performance indicators and benchmarked against others by the regulator or simply by public opinion and the media. Of course it is important to distinguish between efficiency and its perception by customers: a proactive attitude of employees, mainly of those who are in the field and in contact with customers, may lead to a public image of the utility that is better than the reality, or vice versa. Customers value sympathy and diligence and these do not cost more for the water utility: it is essentially a matter of organization, motivation and attitude.

This chapter is dedicated to the improvement of water utility efficiency, with a very substantial emphasis on NRW reduction which, in many cases, is the major symptom of inefficiency. Treatment of symptoms without curing the underlying diseases may lead to short term relief but does not solve the real problems. In water utilities, these problems may be very deep and serious and require a complete turnover of the organization or just minor adjustments of mostly a managerial and technical nature.

As far as NRW is concerned, it may be the symptom of seven "diseases" of the water utility, each one having its own therapy, consisting of a number of measures that contribute to NRW reduction. The measures have different costs and benefits and it is important to find, in each case and on a certain date, the optimum combination of measures to achieve a predefined final result. This is a major challenge for the top management of any water utility.

## 3.2 SETTING EFFICIENCY TARGETS

According to the IWA Efficient Water Management Specialist Group, *"utility efficiency covers a broad range of technical, managerial, financial and governance issues that enable it to perform efficiently and deliver effective services, namely in the following six areas: corporate governance, human resources, accountability towards customers, financial, commercial and technical"*.

This IWA Specialist Group has been very effective in producing manuals of good practice, specialized papers and reports, and training courses on the broad water utility efficiency issue. In particular, it has produced a self-assessment tool for evaluating efficiency in water utilities that can be downloaded from their internet site. The following indicators are used in the IWA Water Utility Efficiency Assessment Matrix (IWA Water Management Specialist Group, 2013):

- Governance Indicators
  - G1 – Mission statement
  - G2 – Strategy
  - G3 – Board of directors and selection of chairperson
  - G4 – Code of conduct and internal accountability
  - G5 – Procedures and quality control
  - G6 – Procurement
  - G7 – Public relations
  - G8 – Professional bodies
  - G9 – Accountability to environment
- Human Resources Indicators
  - H1 – Internal communications
  - H2 – Recruitment and staffing levels
  - H3 – Remuneration levels
  - H4 – Staff training and education programs
- Customers Indicators
  - A1 – Coverage
  - A2 – Deliver of service
  - A3 – Water quality
  - A4 – Customer communication
  - A5 – Customer services
  - A6 – Complaints management
- Financial Indicators
  - F1 – Existence of financial operating projections
  - F2 – Existence of capital planning and projections (CAPEX)
  - F3 – Accounting principles and procedures
  - F4 – Financial reporting and monitoring
  - F5 – External Auditing and annual financial records
  - F6 – Creditworthiness
  - F7 – Tariff setting policies and mechanism
  - F8 – Revenue sufficiency
- Commercial Indicators
  - C1 – Customer information
  - C2 – Meter reading and billing
  - C3 – Collection efficiency: domestic and industrial clients

  - C4 – Collection efficiency: public authorities and bonded clients
  - C5 – Customer metering
- Technical Indicators
  - T1 – Water safety plan
  - T2 – Strategic asset management
  - T3 – NRW management
  - T4 – Operational efficiency
  - T5 – Maintenance
  - T6 – Water resources management
  - T7 – Drinking water quality monitoring

The introduction of these performance indicators, which cover virtually every area of the water utility, is certainly an important measure to introduce efficiency gains and their monitoring over time may contribute to continuous improvement. The PNQS (Brazilian National Plan for Quality in Sanitation) case study presented in this book (CNQA, 2013), shows how the implementation of management programs based on these sort of indicators may introduce exceptional results in water utilities at a national level (in the case of Brazil, due to its very large size, this practically means at continental level). The adoption of indicators means a change in attitude and a focus on excellence that can trigger a deep alteration of the whole water utility. However, the information required to calculate these indicators is not available in the majority of water utilities around the world and, in practice, efficiency can be evaluated initially through a smaller number of simpler key indicators, namely NRW, bill collection, labour productivity (Marin, 2009) and energy consumption.

Target setting is always a complex site-specific task. Objectives may be simply set externally by the regulator, making the job easier for the water utility management in this respect.

As far as NRW is concerned, from the regulator, the owners, the public, the customers and the environment protection points of view, the lower the NRW, the better. However, NRW reduction costs money and water utilities must seek a balance between these costs and the economic benefits of the saved water.

For each water company, at a certain time and for a set of values for number of variables (which will be specified later) there is an ELL – Economic Level of Loss-that corresponds to the minimum NRW value that makes economic sense, namely for which the marginal cost of reducing one cubic meter of NRW is lower than the corresponding marginal benefit. In simple terms, when the ELL is reached, any measure that can be taken to further reduce NRW incurs costs that are higher than the value of water saved.

There is no standard methodology to calculate the ELL: it depends very much on the variables that are used in the cost benefit analysis, for instance the period of time, the value of the water acquired or produced, as well as the interest and depreciation rates. In particular, as previously mentioned, there are two different time frames of ELL: short run and long run ELL. Long run means any time longer than the payback period of the investments (financial perspective) or than the estimated service life of the assets (economic perspective). Short run ELL is essentially related to operational measures that do not require significant capital investment, for instance active leakage control to reduce real losses and inspections of customer installations to reduce apparent losses. However, any comprehensive strategy to reduce NRW must be designed for the long run, at least over a minimum period of twenty to thirty years in order to be able to include the economic study of the measures that require capital investment, such as pressure management, implementation of DMAs – District Metered Areas –, pipe replacement or rehabilitation and customer meter replacement.

In some utilities there is a comfortable excess of water available for supply compared to demand, often referred to as "headroom". In cases where this does not happen, it may be necessary to develop new water sources or expand capacity of the water supply system and include them in the cost-benefit analysis. Under these conditions, of course, NRW reduction projects become even more cost effective for the water utility.

There is no doubt that the private costs unequivocally allocated to the water utility operation must be included in the cost-benefit analysis. However, there are also externalities, mostly intangible costs and benefits related to the environmental and social impact of measures that go beyond the limits of water utilities. For instance, these may include landscape alteration due to construction of new dams to produce more water (eventually deferred or even avoided through NRW reduction) or simply costs associated with reduced flows in rivers and ground water abstractions, also mitigated through NRW reduction. Other examples of this type of costs are traffic delays due to construction of new pipes along with dust and sludge in the city associated with these works. The majority of these costs and benefits are intangible and very difficult to evaluate correctly, although, on the other hand, the additional money in customers' pockets resulting from tariff reductions due to efficiency gains is of a social nature, but certainly tangible.

The more recent water legislation, namely from the European Union, determines that projects should be evaluated through full cost recovery criteria, including both environmental and social costs and benefits. In principle, it is not technically difficult to incorporate these costs and benefits in existing methodologies and computer programs for cost/benefit analysis. What is more difficult is evaluating them correctly, and if this is not done, the result may be just summing up costs and benefits of a very different nature, with different time scales and estimated with very different degrees of precision. Under these circumstances, a sensitivity analysis on the input data is very likely to produce a wide variation of possible results, rendering decision making much more difficult and uncertain. If environmental and social costs and benefits related to reducing NRW are included in the calculations, which seems to be the future trend, the ELL can be renamed as SELL – Sustainable Economic Level of Loss-and their values may be substantially different (Pearson Trow, 2005; Tripartite Group, 2012).

However, ELL is only one of the possible targets for NRW: the regulator or the utility board may define different targets (Trow et al., 2012). For instance, the World Bank has proposed a "Simplified Physical Loss Target Matrix" with four categories and different values for developed and developing countries (Figure 3.1, World Bank Institute, 2005). For each category (A, B, C and D) a number of recommendations have been made (Table 3.1). Of course, setting targets is always a very delicate exercise and proposals are inevitably subject to criticism. However, in cases where the water utility is poorly managed, any realistic target is better than no target at all. A goal that is difficult to achieve but which is seen as realistic by the water utility may make the whole difference in terms of future action and results.

There are currently two major trends in terms of setting NRW targets for water utilities. One is the Frontier Approach, followed by Ofwat, the UK regulator, which aims at ranking operators according to their performance and setting targets on a comparative basis. The other is the Prime Drivers approach, still under development by the IWA Water Loss Specialist Group, led by Stuart Trow, which assumes that there are multiple drivers for NRW in water utilities, namely Economic, Political, Technical, Environmental, Social and Legislative. This is essentially a PESTLE (political, economic, social, technical, legislative and environmental) methodology followed for scenario testing.

Three levels of targets should be defined and monitored: (1) strategic (whole utility, annual basis); (2) tactical (distribution zone, monthly basis); and (3) operational (DMA, weekly). These targets should lead to the reduction of the gap between water bought or produced and water sold, and water utilities should be aware of their starting point and limitations (Pearson Trow, 2005).

| TECHNICAL PERFORMANCE CATEGORY | | ILI | PHYSICAL LOSSES (litres/connection/day) when the system is pressured at an average pressure of: | | | | |
|---|---|---|---|---|---|---|---|
| | | | 10m | 20m | 30m | 40m | 50m |
| DEVELOPED COUNTRIES | A | 1 - 2 | | <50 | <75 | <100 | <125 |
| | B | 2 - 4 | | 50 - 100 | 75 - 150 | 100 - 200 | 125 - 250 |
| | C | 4 - 8 | | 100 - 200 | 150 - 300 | 200 - 400 | 250 - 500 |
| | D | >8 | | >200 | >300 | >400 | >500 |
| DEVELOPING COUNTRIES | A | 1 - 4 | <50 | <100 | <150 | <200 | <250 |
| | B | 4 - 8 | 50 - 100 | 100 - 200 | 150 - 300 | 200 - 400 | 250 - 500 |
| | C | 8 - 16 | 100 - 200 | 200 - 400 | 300 - 600 | 400 - 800 | 500 - 1000 |
| | D | >16 | >200 | >400 | >600 | >800 | >1000 |

**Figure 3.1** Simplified physical loss target matrix (Source: World Bank Institute, 2005).

**Table 3.1** World Bank simplified physical loss target matrix recommendations (Source: adapted from World Bank Institute, 2005).

| Category | Description | Recommendations |
|---|---|---|
| A | Further loss reduction may be uneconomic | • Investigate pressure management options<br>• Investigate speed and quality of repairs<br>• Introduce/improve active leakage control<br>• Identify options for improved maintenance<br>• Assess ELL (short, medium and long term) |
| B | Possibilities for further improvement | IDEM A, plus:<br>• Review asset management policy |
| C | Poor leakage management | IDEM B, plus:<br>• Deal with deficiencies in manpower, training and communication |
| D | Very inefficient use of resources, indicative of poor maintenance and system condition in general | IDEM C, plus:<br>• Fundamental peer review of all activities<br>• 5-year plan to achieve next lowest band<br>• Achieve and maintain Category C performance in defined period |

## 3.3 KNOWING AND REDUCING NRW

### 3.3.1 So much knowledge, so much technology, so much water loss

For many years, a great effort has been made by professional and academic organizations to promote new concepts and methodologies to improve efficiency in water companies, particularly in reducing the NRW levels. A special reference should be made here to the IWA Water Loss Task Force, later transformed into the current Water Loss Specialist Group. This group, mainly the small group of founding members, have played an extremely important role in the overall reduction of NRW worldwide. The IWA Water Balance, the four basic measures to reduce real losses, the four basic measures to reduce apparent losses, the ILI and the design of strategies for NRW reduction should be credited to this group.

The congresses, seminars, conferences and workshops that have been organized throughout the world since 1994, along with books (Farley, 2001; Farley Trow, 2003; Farley *et al.*, 2008; Thornton, 2008; Frauendorfer Liemberger, 2010; Wu *et al.*, 2011), specialized publications and consultation missions fostered by their members have certainly created seeds in many parts of the world and helped to create followers, leading to important local improvements (Lambert and Morrison, 1996; Alegre *et al.*, 2000; Lambert Hirner, 2000; Brothers, 2003; Lambert, 2003; Fanner, 2004; Liemberger, 2006; Rizzo *et al.*, 2004; Alegre Almeida, 2009; Lambert *et al.*, 2014).

However, despite these efforts and the immense developments in knowledge and technology which have occurred, it is clear that NRW is still not under control in many water utilities around the world, whilst in public opinion and the media, the water sector is certainly considered to be very inefficient. This means that NRW is not primarily a technical issue: it depends also on economic, social and political aspects.

There are published estimates of NRW for many water utilities, covering different countries, regions and continents over the globe. These estimates, however, have very different degrees of reliability and where it would be more important to have good data, namely in poorly performing water utilities, the data is invariably missing or not independently audited. However, the order of magnitude of NRW values that exist in many poorly managed utilities is certainly unacceptable. They often do not know their volumes of NRW, but experience shows that water utilities which do not control their NRW have percentages of water loss above 50%, that is, less than half of the water that is abstracted and treated is in fact billed to customers.

Table 3.2 outlines a World Bank estimate of the global volume of NRW. The corresponding values should be regarded only as orders of magnitude as the basic data is mostly estimated and it is certainly not audited. These are average values, which include the larger, better managed water utilities, with much lower values of NRW and a large number of poorly managed utilities with much higher values.

**Table 3.2** Estimates of NRW volumes worldwide (Kingdom *et al.*, 2006; World Bank Institute, 2005).

| | Level of NRW (% of system input) | Ratio | | Volume (billions $m^3$/year) | | |
|---|---|---|---|---|---|---|
| | | Real losses (%) | Apparent losses (%) | Real losses | Apparent losses | Total NRW |
| Developed countries | 15 | 80 | 20 | 9.8 | 2.4 | 12.2 |
| Eurasia | 30 | 70 | 30 | 6.8 | 2.9 | 9.7 |
| Developing countries | 35 | 60 | 40 | 16.1 | 10.6 | 26.7 |
| **TOTAL** | | | | **32.7** | **15.9** | **48.6** |

NRW is one of the major issues which affects the ability of water utilities to remain economically sustainable and, consequently, to finance the expansion of services and/or operational and maintenance activities of the infrastructure. A great proportion of water losses takes place in developing countries, where utilities need additional revenues to finance the expansion of services and most customers suffer from intermittent supply and poor water quality. However, even in the so called developed world, there still are many examples of poorly managed and inefficient water utilities requiring restructuring or extensive efficiency programs, mainly to reduce NRW.

### 3.3.2 The IWA Water Balance in the NRW reduction processes

Any initiative to control NRW requires in depth knowledge of the situation through the quantification of its components and conversion of volumes of lost water (including lost revenue water) to monetary values. One of the cornerstones of an NRW reduction strategy is to understand the relative importance of every component, ensuring that each one is measured or estimated as accurately as possible so that priorities can be established through a series of action plans.

NRW and its components can be determined through a water balance, based on measurement or estimation of water produced, imported, exported, consumed or lost. In the past, there were various formats and settings for water balance calculation and it was virtually impossible to compare data from different utilities. The IWA Water Loss Task Force developed a methodology for calculating the water balance, described in Chapter 2 above, which received wide international acceptance (Alegre *et al.*, 2000; Lambert Hirner, 2000). NRW includes three major components: real losses, apparent losses and unbilled authorized consumption. Real losses correspond to physical loss of water from the distribution system, including leakage and overflow before the customer meter. Typically, these losses contribute to the greatest volume of NRW in most utilities. High real losses mean that a utility has to extract, treat and transport greater volumes of water and that the infrastructure may be oversized. These losses inflate the water utility's production costs and stress water resources since they represent water that is extracted and treated, yet never reaches customers. Apparent losses represent water that is delivered to the customer but not recorded and billed, mostly due to metering errors and theft. These losses should be valued at the selling price of the water services, several times higher than the production cost of treated water and significantly reducing utility revenue. The third NRW component is unbilled authorized consumption, either metered or not metered.

In developed countries, this component is usually very low and related to fire-fighting and in some cases also to irrigation and washing public spaces. However, mostly in developing countries, huge quantities of water used in poor areas are also classified as social authorized consumption: in some cases this water is partly subsidized using taxpayer money, in others it is simply paid by the other customers in their tariffs, often leading to lack of sustainability of the systems.

Essentially, a NRW reduction process starts with an initial water balance and leads to a future situation in which NRW can also be detailed through a new water balance. There are many different prospective water balances for a given utility and many different ways to reach each of them. The future situation that has in fact occurred will become the initial one for yet another phase of the NRW reduction process.

The water balance is crucial for characterizing any stage of these processes and in fact NRW is described, at each stage, through its components (Figure 3.2).

### 3.3.3 Knowing and reducing real losses

There are three main components of real losses: leakage from transmission and distribution mains; leakage and overflows from reservoirs and storage tanks; and leakage on service connections to the customer meter. In many cases, leakage is higher in the service connections than in the distribution mains. Therefore it doesn't make sense to replace only the mains, neglecting connections; actually, replacing old service connections whenever they are found to be leaking, is usually a cost effective measure. Frequently, a part of

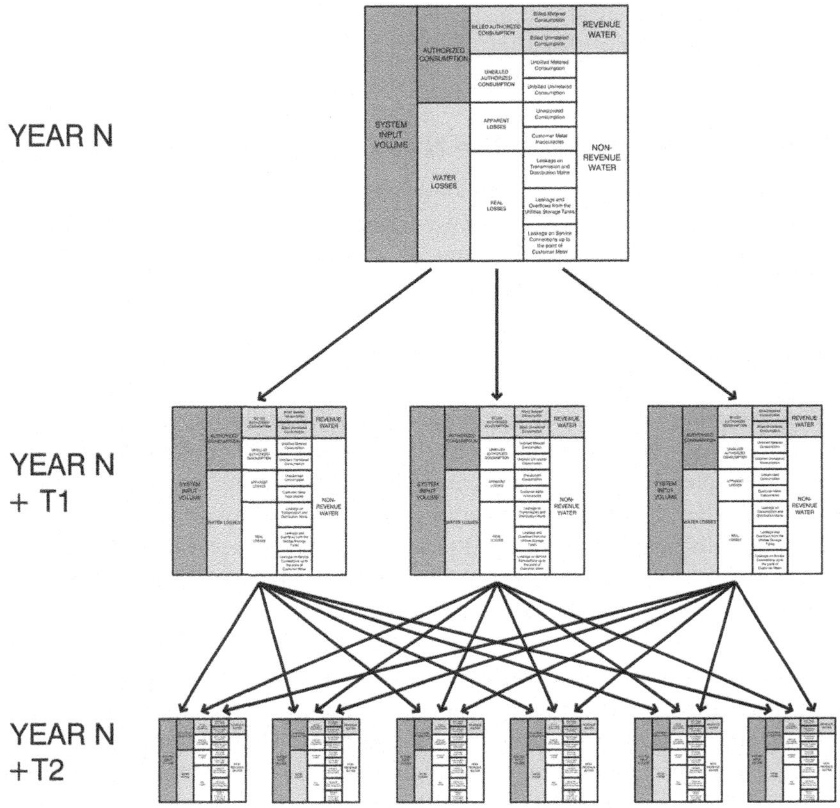


**Figure 3.2** The starting situation and different future NRW targets (one water balance per target; there are different alternatives to reach any future target).

this type of leakage is visible to either the public or utility staff in the street, so they are easy to detect and can be repaired relatively quickly. The other part, however, corresponds to the many invisible small leaks that do not come to the surface and may lead to a great volume of water loss.

The annual volume of losses from all types of breaks, bursts and overflows depends largely on the distribution network characteristics and the leak detection and repair policy practiced by the company. Factors that influence real losses include:

- Age, state and material of pipes;
- Burst frequency;
- Average service pressure;
- Number and length of connections;
- Network Length;
- Soil type, which influences visibility of leaks and bursts;
- Percentage of time that the system is pressurized (a very important factor in areas with intermittent supply).

Reported bursts usually have short combined times for detection, location and repair, and therefore the amount of water lost is relatively small. However, unreported bursts, which can only be detected through active leakage control, may last for much longer periods. Although the leakage rate may be lower than the reported bursts, the corresponding amount of water wasted can be much higher. A strategy to reduce this type of losses, according to the IWA Water Loss Task Force methodology, consists of four main components, as shown in Figure 3.3. The first three strategy measures are mainly management issues,

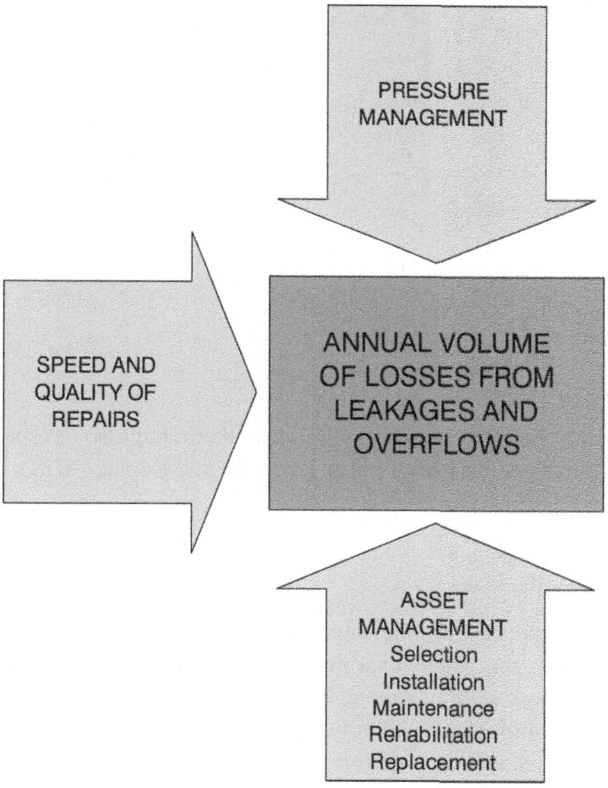


**Figure 3.3** Four pillars of real losses management (Source: IWA Loss Task Force, 2008).

without major investment, which may result in a significant reduction in real losses when the current value is very high. Replacing mains is the least cost effective of the four basic measures to reduce NRW. Pressure management of the water distribution network is likely to produce impressively fast results in terms of real loss reduction and also significantly reduces the pipe burst rate, especially in old networks. Another pillar of real loss reduction is active leakage control. The purpose is to identify invisible leaks, so they can be located and repaired as quickly as possible. This is done by leak location teams who monitor the network using mostly acoustic devices to detect the sound of water escaping from the pipes through holes and cracks (Figure 3.4).

**Figure 3.4** Leak location with geophone.

To reduce the real losses it is necessary not only to detect and locate them, but also to repair them in a timely and efficient manner, based on the concept of Awareness, Location and Repair (ALR):

- Awareness time – time required for the utility to become aware of the leak;
- Location time – time required to locate the leak;
- Repair time – time required to repair the leak.

The volume of water lost will continue to increase until the water utility is aware of a leak, is able to locate it, undertake repairs under pressure or shuts the water for a more extensive repair. An existing invisible leak could run for several months or even years without anyone being aware that it exists. A new significant invisible leak, however, should be detected almost immediately, through the real time monitoring system. The systematic detection of small leaks, which is important to reduce NRW to very low levels, requires the division of the network in DMAs – District Metered Areas – discussed below.

Asset management essentially means good engineering and business practices, including all aspects of a utility's infrastructure management and operation. This requires priority setting and decisions on whether to repair, replace, rehabilitate or leave assets as they are. Therefore, asset management is a systematic approach to managing capital assets in order to minimize costs over their useful life while maintaining adequate service to customers. The reduction of real losses may be enhanced through the following major actions and measures: division of the network into DMAs, analysis of minimum night flows, pressure management and rehabilitation policies.

### *3.3.3.1 Division of the network into DMAs*

Frequently, water utilities do not have reliable data on the volume of water entering the distribution system or on the volume of water supplied to customers, creating difficult conditions for identifying and detecting leaks. A possible solution is to segment the network into smaller areas, the aforementioned DMAs, and to install flow meters at every input and output point of each DMA (Figure 3.5). This division of large distribution zones into smaller areas for leakage control has been successfully used by many water utilities of the world since they were introduced for the first time in 1982 in the city of Plymouth (Farley, 2014. Personal comunication).

The volume of water lost within each DMA is assessed mainly through analysis of night flows and analysis of the flow evolution and tendency over several days or weeks in the DMAs, contributing significantly to the awareness of new bursts and leaks (Morrison *et al.*, 2002; Morrison, 2004). The size of a typical DMA is usually between 500 and 5,000 connections.

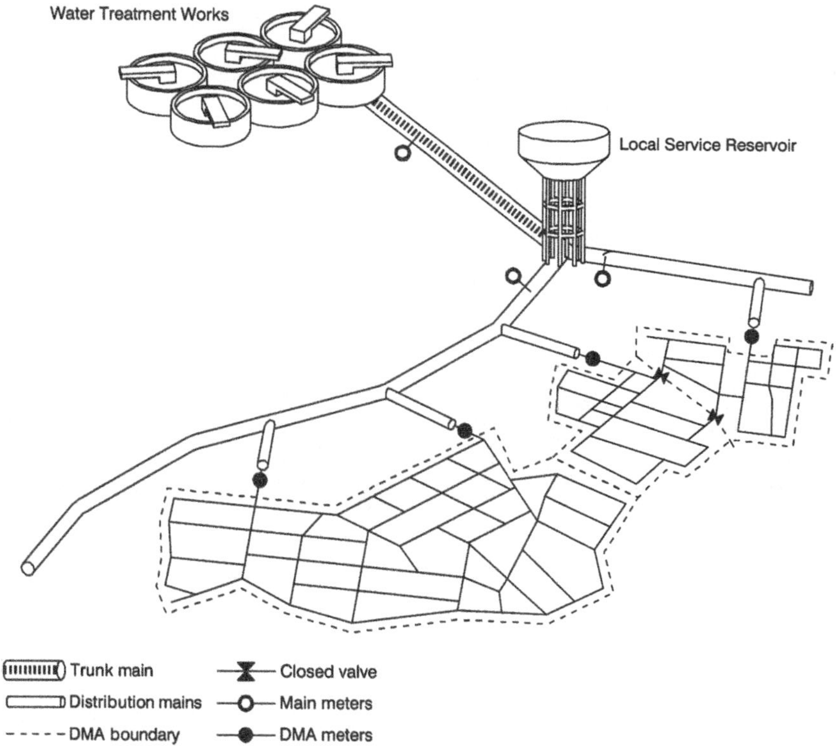


**Figure 3.5** Division of distribution network into DMAs.

The physical division of networks into small DMAs also has potential disadvantages, for instance in cases of fire or bursts when water may not reach the area through alternative pathways and therefore may create reduced flows from hydrants, and it will be more complex to minimize supply interruption to the minimum possible number of customers whenever there is a burst (Farley, 2010).

Instead of a physical division, with one way flow meters at the entry points and no open connections to neighbouring DMAs, it is possible to define bidirectional DMAs installing two-way flow meters at every pipe that may link a DMA to adjacent ones (Figure 3.6). Equipment used to monitor and control

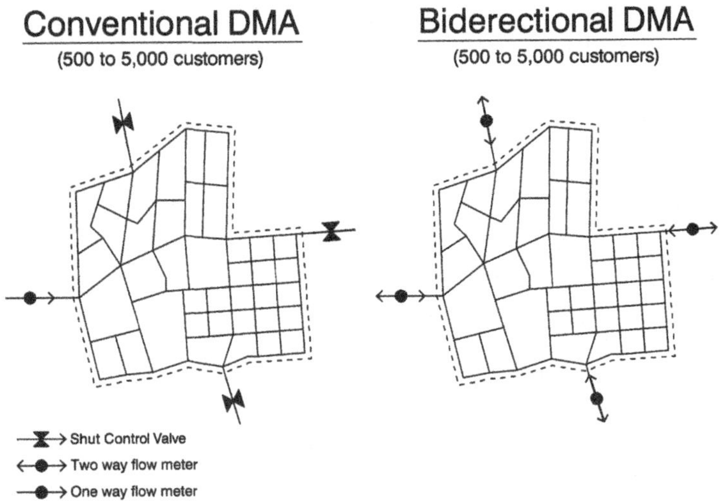


**Figure 3.6** Conventional and bidirectional DMAs.

DMAs include flow meters, section valves, Pressure Reducing Valves (PRVs) and data loggers (Figure 3.7) Usually, this equipment is connected to a telemetry system.

**Figure 3.7** Examples of flow metering and control equipment.

In addition to the establishment of bidirectional DMAs, there are also other alternatives to monitor leakage and bursts, namely:

- The permanent or temporary installation of large numbers of noise loggers over the network;
- The installation of permanent multi-parameter (noise, flow and pressure) monitoring; and
- Using network models to predict leakage.

As the price of noise loggers decreases, it makes economic sense to install them in large numbers over the network and collect data stored locally on a daily basis. Multi-parameter monitoring, originally developed by Martinek Water Management, is a technology comprising of a sensor that can simultaneously measure, record and communicate bidirectional flow, pressure and noise. These sensors are installed at key positions to create 'virtual zones' and allow for monitoring leakage without creating zone boundaries. The three parameters are continuously monitored, measured and analysed during the night hours and compared against a reference value. If a reference boundary value is breached - increased flow, changed flow direction, decreased pressure and/or higher noise-an alarm is created (Farley, 2010).

It is possible to use hydraulic models to predict network areas where leakage is more likely to occur. Essentially the models compare predicted and real values of the pressure, identifying the most likely leakage locations (Bentley Systems).

### 3.3.3.2 *Night flow analysis*

The volume of water lost within each distribution zone or DMA can be assessed in real time through analysis of flow behaviour during the hours of lower consumption, which typically occur at night (Figure 3.8). From the interpretation of the minimum night flow (MNF) it is possible to estimate the level of leakage and additionally to split this estimate into background and burst volumes. A sudden and significant variation of the water volume into the DMA during the MNF period can be an indication of a burst. Leakage is suspected if the MNF rate is greater than a previously measured level.

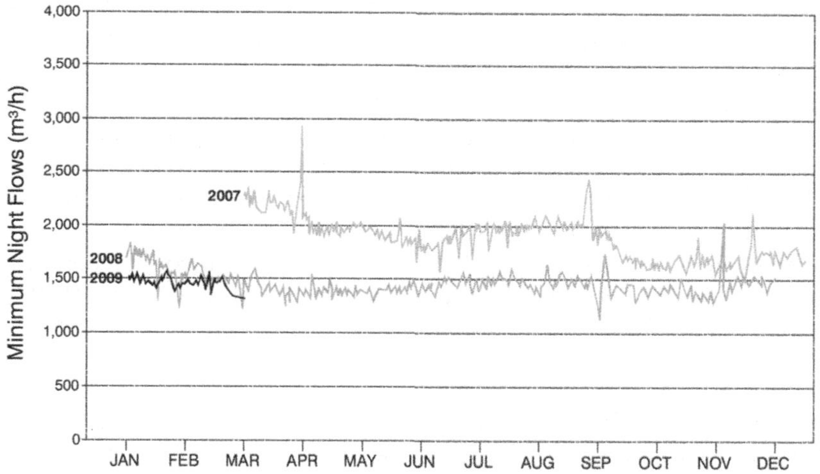


**Figure 3.8** Minimum Night Flows.

### 3.3.3.3 *Pressure management*

Pressure management is one of the key elements of any NRW reduction strategy (Lambert, 2001; McKenzie Wegelin, 2002; Thornton, 2003; Thornton Lambert, 2005 and 2006). During the night period, network pressure increases due to lower flow rates, velocities and head losses in the distribution pipes. It is, therefore, possible to reduce pressure levels at night more than during the daytime. The main benefits of pressure management are a reduction in the leakage rate from existing breaks and cracks as well as in the number of new bursts.

There are a number of methods for reducing pressure, including variable speed pump controllers and break pressure tanks. However, the most common and cost effective is the automatic pressure reducing valve (PRV). PRVs are installed at strategic points in the network to reduce or maintain network pressure at a set level. PRVs maintain the pre-set downstream pressure regardless of the upstream pressure or flow-rate fluctuations. Pressure reduction is particularly important to reduce small, undetectable leaks in joints and connections. The occurrence of high pressures and significant daily fluctuations are the principal causes of infrastructure deterioration and burst rate increase.

It is surprising that, in many water utilities, the strategy for pressure management does not include what should appear to be an obvious step: the analysis of the existing networks, looking for opportunities to alter the boundaries of the distribution zones and eventually to replace pumping with gravity flow. This relates to altering the existing network towards an optimized situation, taking into consideration local topography and urban pattern, namely the number of building floors and existence or not of individual storage tanks.

Pressure management corresponds to controlling pressure at any point during the 24 hour daily cycle and every day of the year, keeping it at the lowest value that does not interfere with normal water use by customers. This means that it is necessary to reduce the pressure more substantially during the night than during the day. The PRVs may be programmed to reflect and respond to the flow variations either directly, using telemetry and remote control, or indirectly, using simple timers, when consumption patterns are predictable.

In either case, the continuous action on the networks to stabilize pressure also has disadvantages: mains are subject to sudden pressure variations which may lead to more bursts in weak older pipes whilst management of the entire network becomes more complex, requiring a higher level of expertise and in-house capacity to maintain and repair the more sophisticated telemetry and communications systems necessary to support real time control. This real time control has been enhanced by artificial intelligence programs that utilize the great number of time series flow and pressure data in critical points of the networks, and give automatic instructions to open or close control valves and to alter variable speed pump settings. It is obvious that this type of control over the networks is possible in the best utilities, with high levels of monitoring and instrumentation, but it may not be possible in the majority of water utilities, where the technology and, especially, the technical expertise, simply do not exist. In these cases, it may be wiser to propose simpler and more robust methods of pressure management not requiring real time control and which are less technology dependent, for instance PRVs regulated by timers.

If pipes and fittings are of good materials and relatively new and if the construction was also good, the impact of pressure variations on bursts and leaks would be lower: in such networks, NRW is less pressure dependant.

### *3.3.3.4 Rehabilitation policies*

A high volume of real losses means that the network is not in a good condition. The deterioration of pipes and other components is reflected in the frequency of bursts and leakages (Savic *et al.*, 2005). Differences in construction and maintenance practices and in the selection of pipe materials contribute greatly to these differences.

There is an ongoing dilemma between rehabilitating or replacing old pipes (Parker, 2009). A proper maintenance regime extends the useful life of infrastructure but, at some point, bursts may become too frequent to be acceptable by the affected customers, whilst, on the other hand, repair costs become so large that replacement becomes an unavoidable option. Replacing pipes is expensive and, in urban areas, it creates traffic disturbance and congestion. However, as an alternative to traditional methods, the trenchless installation, replacement and rehabilitation of pipes is becoming more common. In fact, although trenchless technologies have been available for decades, the costs were too high in comparison with traditional methods. This situation has changed and currently there are many situations in which trenchless options result in more economic solutions, in addition to the obvious advantages to city life. The fixed costs of the specialized equipment and manpower involved in trenchless installation and rehabilitation of pipes are very high and the solution is competitive only where there is a mature market that ensures that they are in operation most of the time. This is essentially the case in the more densely populated urban areas of developed countries, in which digging deep trenches in busy streets just to replace or rehabilitate a comparatively small pipe simply does not make sense and may not even be allowed.

There is a very important difference between replacing an existing pipe and installing a new one using trenchless technologies: in the first case, there is already a clear open space, without obstacles, that can be used, whilst in the second there is always the possibility of unforeseen impediments that make it necessary to go back and attempt an alternative channel for the new pipe in the congested underground of most urban streets. Usually, water distribution pipes are installed at a low depth in pathways and pavements, with building connections relatively close to each other. Under these circumstances, the trenchless option is less applicable, except to overcome street junctions (Figure 3.9).

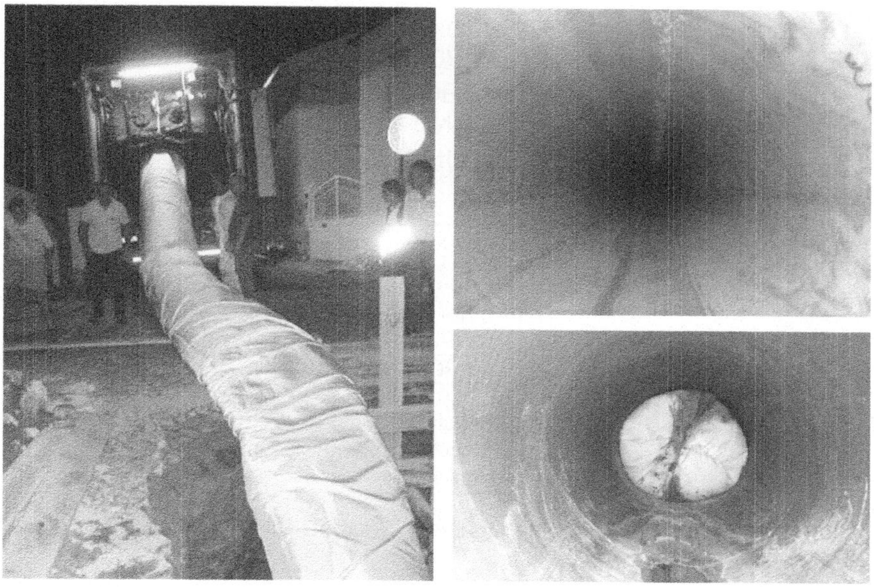

**Figure 3.9** Trenchless rehabilitation of pipe.

### 3.3.4 Knowing and reducing apparent losses

Apparent losses result from the under-registering of customers' water meters and from unauthorized consumption by theft or illicit use. They have a significant financial impact because they are valued at selling prices, typically several times higher than the production costs. In fact, the water bill often includes wastewater service as well and sometimes also solid wastes service, both related to water consumption for billing purposes. Therefore reducing one cubic meter of apparent losses may be worth many times the reduction of one cubic meter of real losses.

The volume of water consumed but not billed to customers is determined by four main causes (Figure 3.10).

#### *3.3.4.1 Meter inaccuracy*

Water meters do not precisely measure the volume that actually goes through them due to meter degradation (age or illegal manipulation) or incorrect type, installation or sizing. The vast majority of customer water meters have moving parts that are affected by age, water quality and also by the presence of grit, as a result of a burst, for instance; the older they are, the higher the under-metering errors, which can easily exceed

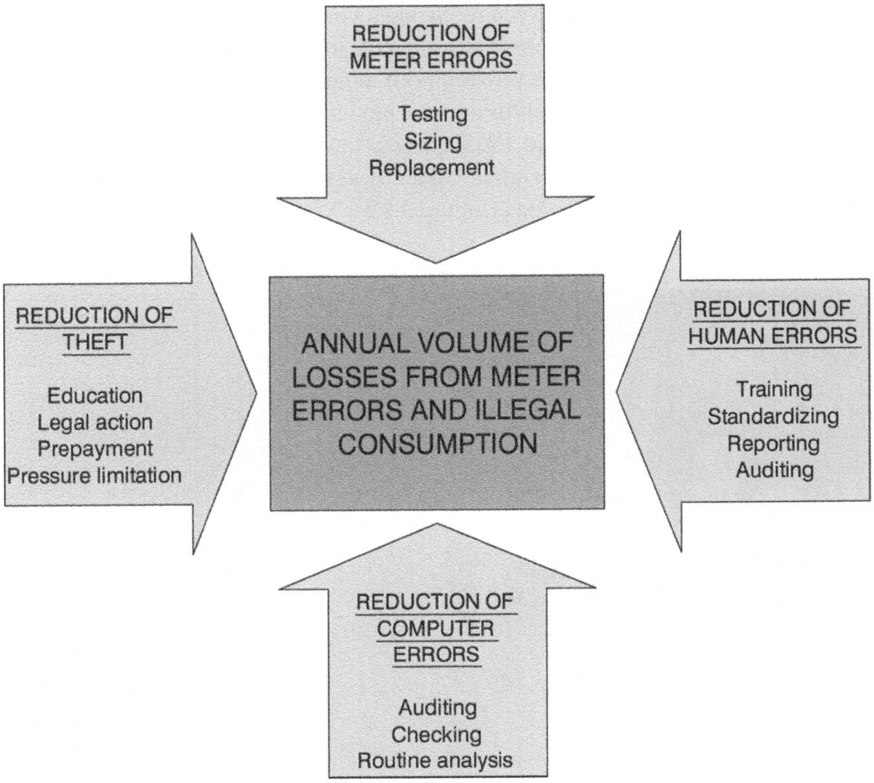


**Figure 3.10** Four pillars of apparent losses management (Source: IWA Water Loss Task Force, 2008).

10% when they are around 10 to 12 years old. Under these circumstances, it makes economic sense to replace the meters when they are 5 to 6 years old.

Water meters without moving parts are already available on the market, based on the same technologies that are used in the larger diameter flow meters and it is very likely that in the future, with the increase of battery duration, they will replace the existing mechanical ones.

It is necessary to ensure that customer meters are in good working order and that equipment is replaced in due time, according to a simple cost-benefit analysis. Large customers deserve special attention: a more specific selection of the meter type and size, more frequent readings and, eventually, the installation of a data logger.

### *3.3.4.2 Human errors in meter reading*

These errors occur mostly during the reading of meters. Some of these errors may be detected in the office, before the bill is issued, through an analysis of previous consumptions. However, any alteration of a real meter reading must be made very carefully: an abnormally high reading may also result from a leak in the customer's house and the longer it takes to detect it, the worse it will be. On the other hand, nearly zero consumption may happen in situations with entirely different meanings: when the customer is not using the house or when the meter wheel has stopped moving due to a fault or to interference. It is important to undertake a systematic control of a customer's historical consumption and a comparison with their neighbours to identify anomalies and trends. Because of the high number of consumers, this analysis must be carried out automatically, using specialized computer programs and eventually expert systems.

It is also necessary to identify every address in the area covered by the water utility and to check that they correspond to customers in the billing system, using Geographical Information Systems.

The use of AMR - Automated Meter Reading - is becoming more generalized as the cost of this equipment is reduced; it has many obvious advantages and virtually eliminates human errors in meter reading.

#### *3.3.4.3 Data handling and accounting errors*

Data handling and accounting errors occur essentially in billing data processing, in most cases being due to the need to estimate consumption because of the lack of real meter readings. These estimates are made based on the historical customer consumption record and corresponding estimation errors are corrected through the next real reading.

A robust customer billing system is essential to avoid errors and complaints. There are a number of proprietary systems available on the market to water companies as external services. Less frequently, water utilities develop and maintain their own systems, but this option requires a strong development team to deal with its maintenance. These programs are apparently simple, but the management of the great number of possible exceptions is particularly complex and time consuming. The best billing software packages have built-in analysis functions that can identify potential data handling errors and report them for verification.

#### *3.3.4.4 Unauthorized consumption*

There are always people who manage to steal water through illegal connections or by tampering with meters. The definition of a strategy to reduce apparent losses must include data mining in the customer database for abnormal consumptions (total and patterns). It is also important to have historical consumption data for each customer, if possible for several years, and to read meters frequently, at least twice a year, to detect hidden leaks in customer installations, faulty meters and water thefts.

Unauthorized consumption may be responsible for a significant loss of revenue of water utilities. It is also essential to detect and eliminate theft cases to avoid the propagation of these bad examples. The combat process starts inside the water company through computer analysis of the customer database. This operation consists essentially of the detection of an anomalous consumption and of addresses that are suspiciously not included in the billing system.

Customer meter readers who visit installations a number of times a year and the installation of AMR – Automated Meter Reading-are important to identify those situations.

### 3.3.5 Knowing about and reducing unbilled authorized consumption

Ideally, all water that is consumed should be billed; otherwise it will be indirectly charged to the paying consumers through their tariffs, and it is an incentive for wastage and unnecessary consumption.

In any case, except for the water that fire brigades take for fire fighting from specific hydrants (which is very easy to estimate with sufficient accuracy) all consumption should at least be measured to prevent wastage and for accountability purposes, including control of the NRW.

However, in many water utilities there are political decisions that determine that certain types of consumption and customers are not charged for water services. This consumption may be metered or un-metered, but in each case, the bills are not issued or they are issued with a zero tariff.

Overall, it is a management responsibility to attempt to eliminate situations of inequity as much as possible, measuring or at least estimating as accurately as possible every volume of water consumption.

There are different points of view concerning the meaning of universal access to water and water services. It is clear that water is essential to life and without water an urban house becomes uninhabitable. However, it is better to subsidize directly through the social security system those customers who cannot afford to pay for water services (or for other essential goods and services), so that the water bills are issued in the normal way to everybody and actually paid for; it is more transparent to everybody and it contributes to avoiding wastage.

## 3.4 DEVELOPING A STRATEGY TO REDUCE NRW

The starting point to reducing NRW in a water company is evaluating the magnitude of the problem and its context. This is always a site specific issue and the evaluation may be attempted by an internal team in the water utility, eventually assisted by a consultant. In Malcolm Farley and Stuart Trow's book on Water Loss Reduction (Farley and Trow, 2003), an excellent summary of a standard strategy to reduce NRW is presented (Figure 3.11, reproduced with the permission of the authors).

The quantification and characterization of NRW are best achieved through the preparation of the company Water Balance using the IWA methodology. This global water balance gives a first order of magnitude of the quantity and value of the water that is being lost. Its components give a preliminary indication of where and why water is being lost and gives general indications on priority actions.

However, to understand the problem further, it is necessary to undertake water balances for each distribution zone. It is also necessary to develop relatively lengthy audits in various departments of the company and in to its infrastructure and operational practices, requiring field visits and personal interviews. The strategy objectives, in particular, depend very much on the level at which the action is going to take place: technical, middle management or top management. The higher the level, the more ambitious the initial objectives should be. These objectives may be reevaluated at a later stage but what should not happen is setting objectives that require levels of decision, authority, time or resources that are not guaranteed.

From personal experience, four to six years are necessary to develop and implement an NRW reduction strategy leading to excellence in previously poorly performing water companies, but there are almost always many things that can be done to produce results in less than a year.

Establishing a certain percentage of NRW as an objective to reach is certainly not the best option from a technical perspective, but it is not uncommon that the top management or the regulator state that the water company should have a water loss lower than 20% within three years, for example. It then becomes a technical challenge to demonstrate that the objective is or is not possible, how much it will cost to achieve it and what should be done to do so.

Whichever way objectives have been set, an NRW reduction strategy invariably involves moving from the current situation, with known values of SIV and RW, through reduction of the SIV (acting on real losses) and the increase of the RW (acting against apparent losses and authorized not billed consumptions).

In the example presented in Figure 3.12, the initial values of the SIV and RW are respectively 104,000 $m^3$/day and 48,000 $m^3$/day. For each set of values of SIV and RW, it is possible to calculate the corresponding impact on the annual company results. These additional annual results may be compared with the annual costs of the measures used to achieve them and, therefore, to evaluate the corresponding cost effectiveness. From this figure, it is clear that simply establishing a certain percentage of NRW as the final objective may correspond to completely different final values of SIV and RW. Alternatively, each set of SIV/RW values may be reached in a number of ways, certainly with different measures and different costs.

In particular, it is possible to start moving horizontally first (prioritizing the reduction of apparent losses), vertically (priority to the reduction of real losses) or horizontally/vertically in succession (priority to both).

| QUESTION/SOLUTION | TASK |
|---|---|
| **1.** HOW MUCH WATER IS BEING LOST?<br><br>- Measure components | WATER BALANCE<br><br>- Improved estimation/measurement techniques<br>- Meter calibration policy<br>- Meter checks<br>- Identify improvements to recording procedures |
| **2.** WHERE IS IT BEING LOST FROM?<br><br>- Quantify leakage<br>- Quantify apparent losses | NETWORK AUDIT<br><br>-Leakage studies (reservoirs, transmission mains, distribution network)<br>-Operational/customer investigations |
| **3.** WHY IS IT BEING LOST?<br><br>- Conduct network and operational audit | REVIEW NETWORK PRACTICES<br>Investigate:<br>- historical reasons<br>- poor practice<br>- quality management procedure<br>- poor materials/infrastructure<br>- local/political influences<br>- cultural/social/financial<br>- factors |
| **4.** HOW TO IMPROVE PERFORMANCE?<br><br>- Upgrade the network<br>- Design a strategy and action plans | STRATEGY DEVELOPMENT<br><br>- Update records systems<br>- Introduce zoning<br>- Introduce leakage monitoring<br>- Address causes of apparent losses<br>- Initiate leak detection/repair policy<br>- Design short/medium/long term action plans |
| **5.** HOW TO SUSTAIN PERFORMANCE?<br><br>- Ensure sustainability with appropriate staffing and organisational structures | TRAINING AND O&M<br>Training:<br>- improve awareness<br>- increase motivation<br>- transfer skills<br>- introduce best practice/technology<br><br>O&M:<br>- Community involvement<br>- Water conservation/demand management programmes<br>- Monitor action plan recommendations<br>- Introduce O&M procedures |

**Figure 3.11** Key questions, solutions and tasks for NRW reduction (Source: Farley and Trow, 2003).

These options have very different impacts on customers, on the infrastructure, on costs, on the benefits and on the development of the strategy itself.

In a water utility there are often very different values of NRW in the various distribution zones; therefore, it is not possible to design an NRW reduction strategy that does not take such differences into

## Additional Annual Results

(million euros)

| SYSTEM INPUT VOLUME ($m^3$/day) | REVENUE WATER ($m^3$/day) 48,000 | 50,000 | 52,000 | 55,000 |
|---|---|---|---|---|
| 104,000 | NRW 54% 0.0 | 1.1 | 2.2 | 3.8 |
| 86,000 | 2.0 | 3.1 42% | 4.2 | 5.8 |
| 80,000 | 2.7 | 3.8 | 4.9 35% | 6.5 31% |
| 75,000 | 36% 3.2 | 4.3 | 5.4 | 7.1 |
| 70,000 | 3.8 | 4.9 29% | 6.0 | 7.6 |
| 65,000 | 4.4 | 5.4 | 6.5 20% | 8.2 15% |
| 60,000 | 4.9 | 6.0 | 7.1 | 8.7 |
| 56,000 | 5.4 | 6.5 | 7.5 | 9.2 |

**Figure 3.12** Evolution of SIV and RW over time and impact on annual results.

consideration: it is essential to undertake a water balance for each individual distribution zone and to develop a separate sub-strategy for each one.

The global water balance uses total values, namely SIV and RW, which are usually audited and incorporated into the annual company accounts. These total values are essential for filling the information gaps and calibrating distribution zone water balances, as the company water balance should be the sum of the water balances of all distribution zones into which the network is divided.

In any case, in the initial stages of the NRW reduction process in a water utility it is already very important to have a global water balance and a high level global strategy. This global strategy should include the current and target water balances, the main measures and actions that are required to reach that target and respective costs, and a preliminary investment plan. It could always be possible to obtain more detailed information, but this would not necessarily lead to more precise results and the opportunity for action might be lost. Indeed, to obtain and validate the additional information required to define a more detailed initial strategy, it would be necessary to undertake water audits in the distribution zones, including field visits and to develop relatively lengthy audits in the various departments of the company, its infrastructure and operational practices.

The reduction of each NRW component, according to the IWA methodology mentioned above, can be performed through a number of specific actions (Table 3.3).

In addition to these specific actions, there are others that are common to other aspects of a water utility organization, such as the creation of a Central Control Room, a Telemetry System and a corporate GIS.

NRW reduction is a long-term process and, therefore, the corresponding strategy must cover at least a twenty to thirty year period. Of course, it must be divided into shorter periods for evaluation and budgeting purposes. It is necessary to ensure that all NRW components are addressed in the strategy and ensure that it is feasible in terms of workload and budget. It is also important to realize that any strategy for NRW reduction must include repetition and continuous pressure on the various action fronts, otherwise a return to poor values is very likely.

**Table 3.3** Components of NRW and measures to reduce them.

| Components of NRW | Measures |
|---|---|
| Unbilled Metered Consumption | • Bill every consumption |
| Unbilled Unmetered Consumption | • Install meters and bill every consumption |
| Unauthorized Consumption | • Analysis of historical consumptions<br>• Local inspections<br>• Legal actions<br>• Interruption of customer's supply |
| Customer Meter Inaccuracies | • Local inspection of the meter<br>• Replacement of the meter |
| Leakage on Transmission and Distribution Mains | • Fast repair of bursts<br>• Active leakage control<br>• Pressure management<br>• Replace or repair the mains |
| Leakage and Overflows at Storage Tanks | • Telemetry<br>• Local inspection for overflows<br>• Repair the tank surfaces |
| Leakage on Service Connections | • Fast repair of bursts<br>• Active leakage control<br>• Pressure management<br>• Replace or repair the connections |

## 3.5 PERFORMANCE INDICATORS OF NRW

Water losses are sometimes expressed as a percentage of the SIV. Despite being simple and powerful to alert the non-technical audience to often shockingly high figures in many water utilities, this performance indicator is misleading since it favours utilities with high consumption, low pressure and intermittent supply and does not even distinguish, for instance, between real and apparent losses. Hence, other indicators should be used for the operational management of NRW and even for communication purposes (Lambert *et al.*, 1999; Alegre *et al.*, 2000; Lambert Hirner, 2000; Lambert McKenzie, 2002; Lambert Liemberger, 2005; Liemberger, 2007; Alegre Almeida, 2009; Lambert *et al.*, 2014). The IWA best practice suggests its use only as a financial performance indicator and states clearly that it is unsuitable for assessing management efficiency of the distribution systems.

Separate indicators are recommended by IWA for real and apparent losses, namely:

- For real losses:
  - litres/service connection/day (or $m^3$/km mains/day if the system has less than 20 service connections per km of mains);
  - ILI - Infrastructure Leakage Index

$$ILI = \frac{CARL}{UARL}$$

CARL - Current Annual Real Losses;
UARL - Unavoidable Annual Real Losses
- UARL (l/d) = 18 x Lm+0.8 x Nc+25 x Lp) x P
- Lm = mains length (km)

- Nc = No. Service connections (mains to property lines)
- Lp = total length of service connection (property line to customer meters)
- P = average pressure (m)

- For apparent losses:
  - % of Authorized Consumption.

The ILI is the most rigorous for real losses because it takes into account the pressure in the network, allowing for meaningful comparisons between utilities (Lambert McKenzie, 2002; Lambert *et al.*, 2014).

An extremely well managed system can have an ILI close to 1.0 (i.e., CARL = UARL), but this should not necessarily be the target unless potable water is either scarce or expensive. An ILI in the range 1.0 to 1.5 represents an excellent technical level of management of real losses, and has now been achieved by more than a hundred utilities worldwide. Figure 3.13 presents ILIs for 72 water utilities in 12 high income European countries, circa 2012. ILIs slightly less than 1.0 can be achieved in low pressure and predominantly small systems (Lambert *et al.*, 2014).

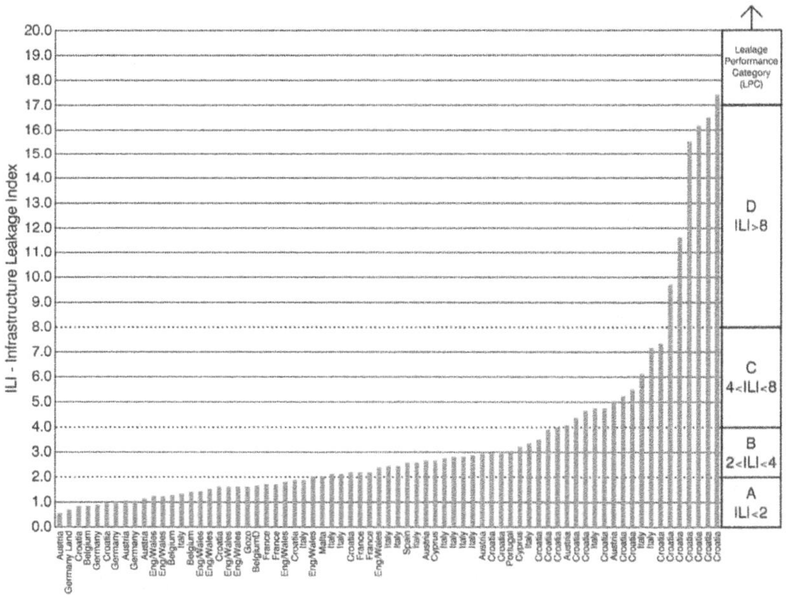


**Figure 3.13** ILIs for water utilities in 12 high income countries circa 2012 (Source: Lambert *et al.*, 2014).

## 3.6 TECHNOLOGIES USED IN WATER LOSS CONTROL

The simplest means of locating leaks is to look for water on the streets or for abnormal moisture, moss formation or vegetation growth, but in addition to fixing visible bursts and leaks, water utilities must take systematic measures to detect the leaks that do not appear at the surface. For this, it is essential to mobilize active leakage control teams who systematically monitor the network using both old and new technologies, especially acoustic devices. In very busy streets, these tasks must be carried out mostly at night, when there is less traffic and background sound interference. The success of acoustic detection depends largely on the pipe material. For example, it is considerably more difficult to find leaks in plastic pipes than in metallic pipes, which have better sound propagation characteristics.

However, this is not the place to describe in detail the existing and emerging technologies for water loss control as a considerable volume of published material exists on this specialist, technical subject (Pilcher *et al.*, 2007; Farley, 2010; Hamilton Charambolous, 2013). New proposals appear frequently and the simplest way to keep up-to-date is to attend any of the specialized congresses, seminars and fairs, reading specialist water loss journals and more importantly, to conduct regular internet searches. There are the so-called old technologies, supported by long-term practices, which have been in use for many years, such as the apparently simple but extremely difficult to master listening stick, whilst occasionally new solutions appear on the market: some stay for a long time and may see improvement developments, whilst others simply disappear. In addition to leak detection technologies, there have been very interesting developments on the repair side, mainly fixing of smaller leaks in pressurized pipes, without interrupting the supply.

The following list attempts to incorporate the major types of leak detection and control methods (adapted from Pilcher *et al.*, 2007; Farley, 2010; Hamilton *et al.*, 2010; Hamilton Charambolous, 2013).

1. Gas injection, in which the pipe is filled with a tracer gas, mainly Hydrogen, Nitrogen or Helium that escapes through the point of the leak and is detected with a probe at the surface;
2. Stethoscope or listening stick;
3. Leak noise correlation, in which sound picked up by sensors deployed at two different locations is automatically analysed by digital correlators;
4. In-pipe acoustic techniques using a noise sensor to locate the position of the leak from inside the pipe:
   a. Tethered systems;
   b. Free swimming systems;
5. Noise loggers – non correlating;
6. Electronic amplified listening devices (electronic versions of the listening stick);
7. Other techniques:
   a. Thermal imaging where water from the leak is of a different temperature than the surrounding ground and can be detected with an infrared thermographic camera;
   b. Ground penetrating radar (observation of wet or disturbed ground or cavities around the pipe);
   c. Optimization for leak location using network models to find areas of leakage;
   d. Step testing: manual and electronic (associated with DMAs);
   e. Multi-parameter monitoring (pressure, flow and noise).
8. Recent proposals in repair technology:
   a. Repair clamp (locally or remotely installed);
   b. Polyurethane lining technique;
   c. Trenchless automatic leak repair (sealing the leak from the inside using chemicals and/or robots).

Figure 3.14 presents the most important technologies for water loss control in chronologic order. It is important to stress that these technologies may be important to reduce NRW faster and more successfully,

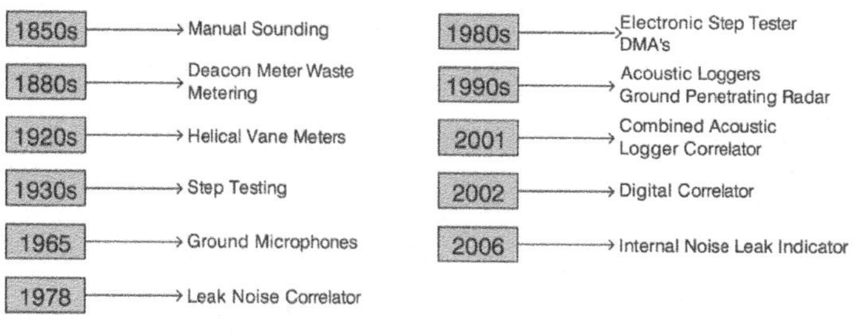


**Figure 3.14** Technologies for water loss control (Source: Pilcher *et al.*, 2007; IWA Water Loss Task Force, 2008).

but in water utilities, as in other organizations, technology is just an accelerator; its effectiveness and success depend on the incorporation into a comprehensive strategy with many other components.

Some of these technologies are very specialized and are offered, at least in the early versions, as outsourced services. Later on, in some cases, the more extensive use and adoption of different marketing strategies may permit the acquisition of equipment and training of water utility collaborators in their use.

## 3.7 TOP AND MIDDLE MANAGEMENT TRAINING

Probably the most effective way of improving performance of an existing water utility is providing training to its top and middle management. This training is better achieved through formal tailor-made programs for each management team, in a restricted and confidential environment, using the water utility's site specific data and problems as a real case study.

The basis of such a specific training program could be the following:

1. Water Supply Systems
   a. The water cycle
   b. The urban water cycle
   c. Water supply systems
   d. Integrated management of the urban water cycle
   e. Effectiveness and efficiency
   f. Performance indicators
   g. Management control
2. Management of Change in Water Companies
   a. Effectiveness, efficiency and excellence
   b. Theoretical models
   c. Restructuring water utilities
   d. Relation with the stakeholders
   e. Case studies
3. Basic hydraulics
   a. Flow equations
   b. Head losses in pipes
   c. Hydraulic gradient
   d. Pressure and pressure management
   e. Pumps
   f. Pressure reducing valves
   g. Hydraulic modelling
4. NRW and NRW Reduction
   a. The water balance
   b. Target setting: ELL and other alternatives
   c. Reducing NRW
      i. Real losses
      ii. Apparent losses
      iii. Authorized unbilled consumption
      iv. Performance based contracts
5. Sustainability
   a. Tariffs, transfers and taxes
   b. Sustainability strategies and models
   c. Full cost recovery

6. Innovation and new technologies
    a. Introducing innovation in water utilities
    b. Existing and emerging technologies
7. Future trends
    a. Water systems of the future
    b. Universal access to water

A very interesting example of successful training programs of this type is the Brazilian PNQS – National Quality Plan in Sanitation – which has been developed systematically at a national level, reaching virtually every water utility in the country. It is impossible to evaluate the impact of programs such as this on global results at a national level, but it certainly has a multiplier effect, through improvement of organizations and most importantly, of individual competence, motivation and attitude.

## 3.8 THE WATLOSS/TOPMAN OPTIMIZATION MODEL

### 3.8.1 Introduction

NRW reduction is at the top of the priority list of management of most water utilities. Defining strategy and targets to be obtained is a complex iterative process in which the decision maker asks many what-if questions to his/her close collaborators and consultants. The author has developed an interactive computer program that may be used directly by a top manager, assisted by a NRW specialist, which ultimately leads to the definition of a detailed strategy which may or may not aim at the ELL. The selection of the strategy period to be covered is an important top management decision unless it is fixed by the regulator. However, in any case, the top manager, in addition to the usual 20 – 30 year period, may want to concentrate also on intermediate results in a set period, for example his/her own mandate.

The computer program developed may be used at any water utility, in any stage of development and addresses the entire process of reduction of NRW, from data collection to the implementation of the strategy. Using some fifty figures that should be available at any water utility, most of them in the official annual reports, it builds the current IWA water balance both in terms of volume and monetary value. The water balance identifies the absolute and relative severity of the components, the "seven diseases" with a common symptom: NRW.

For each of these "diseases" there are one or more measures to be introduced and a certain measure can partially "cure" more than one "disease". Each measure has a unit cost (for example, a 50 Euro per customer water meter replaced) as well as a benefit, which typically follows a law of diminishing returns: if customer meters are simply organized in descending order of age, the benefit of replacing the first meter is likely to be much higher than replacing the last one. However, there are, of course, more sophisticated alternatives to detect the most cost effective replacements. The definition of costs and benefits is crucial to the final result: these values, like most others, are site specific and can be specified by the user.

One of the questions that top management or indeed any customer should ask is: what is the ELL for a particular water utility?

This ELL (the Economic Level of Loss) refers to NRW and not only to real losses as its does acronym, Economic Level of Leakage, much used in the UK, does. However, it is interesting to refer here to the report *"Best Practice Principles in the Economic Level of Leakage Calculation"* (Tripatite Group, 2002), jointly developed by OFWAT, Environmental Agency and Department for Environment, Food & Rural Affairs (DEFRA). This report provides a framework to calculate the Economic Level of Leakage, but their guidelines are only indicative, as there is no single approved method. The report made it clear that calculations should be based on each set of company data, being calculated separately for each distribution zone (not for the whole company), and it should include environmental and social costs and benefits. The calculation of economic levels of leakage has also been addressed in depth in other publications, namely in Pearson Trow (2005), and in the IWA Water Loss Task Force Target Setting Initiative (Trow *et al.*, 2010).

Unlike other approaches, in which the ELL has to be calculated or estimated in advance, the program proposed here calculates it as a result of an optimization algorithm. It is achieved as the result of a set of measures, each with a cost, a benefit and an implementation schedule.

ELL is highly dependent on time: for instance short term ELL is different from long run term ELL although both can be calculated by the program, depending on the input variables. When water companies have a stable institutional structure and are professionally managed and insulated from political cycles, NRW reduction depends mostly on technical and economic issues and the period selected by management to calculate the ELL tends to be longer. In many situations though, the time available to define a strategy and implement the corresponding measures does not exceed one or two mandates (three or six years). Three years is perhaps the minimum period required for successful implementation of a change management process, but six years is a much safer option. In any case, the program can optimize measures to be taken over any period. For instance, if a manager has only one year ahead, he/she may want to know what are the most cost-effective measures or simply the most effective ones to minimize NRW during that period. The judgement on whether this is good or bad for the company is purely for the user: indeed, short term actions must be taken within a long term framework.

Reducing NRW very fast is challenging but it may not be in the best interest of the water utility. It is necessary to take into consideration further evolution of NRW over time and measures that may be taken to keep it under control. In particular, after NRW has reached a certain level, it will tend to increase again if the right measures to sustain the optimised levels are not implemented. It is very difficult to predict the long term cost effectiveness of these measures. For instance there are no simple relations between the average ages of pipes or customer meters, and water loss; however, once a certain level of NRW has been reached, it is possible to select, for the coming years, measures that are more cost effective to maintain or improve NRW performance.

The program may be used at any stage to optimize future measures, taking into consideration the results achieved so far. At the end of the restructuring process, there will be a certain value for NRW. If nothing else is done after that, it is expected that NRW will increase again: for instance breaks in the mains will get bigger, new breaks and bursts will emerge and water meter under registration errors will increase.

To keep NRW under control it is essential and highly cost effective to undertake active leakage control and pressure management; this may be sufficient for some time, but as the average age of the pipes increases, it eventually reaches a certain point, dependent on material, quality of construction and operating pressure, after which the number of bursts and leakage sharply increases and they have to be replaced. For each water utility, it is possible to control average infrastructure age, namely of the pipes, connections and water meters, through a replacement program. This certainly helps to keep NRW under control, but it is virtually impossible to establish an optimum average age, as it depends on many unknown factors, namely the quality of the maintenance (as happens with airplanes, for example), the average pressure, and the pressure variations and transients. In particular, it is expected that pipes last for over a century but nobody can make realistic predictions for this period (earthquakes, floods, wars, technological innovation, etc.). However, it makes sense to replace, for example, at least 1% of mains and house connections and 10% of water meters every year until, eventually, substantially better alternatives appear on the market.

The program proposed allows for the introduction of limitations to the rate at which measures can realistically be undertaken: for example, in a city, it is not possible to replace more than a certain number of kilometres of pipes each year, otherwise the impact on traffic and mobility would be unacceptable. Again, this limitation may be overruled by the user if he/she assumes that it is not a problem or if an option is made for trenchless installation and rehabilitation. It is also possible to introduce limitations on annual investments, for example spreading spending over a longer period. The optimization can be made in a number of different ways and using many combinations of restrictions, leading, of course, to different results, as would happen in a traditional decision making process, by top management, without this program.

### 3.8.2 Basic Data

The most important data required to run the program is presented in Table 3.4.

**Table 3.4** Data required to run the program.

| | Data required | |
|---|---|---|
| System Input Volume ($m^3$/year) | Under metering in customer water meters (%) | Purchase or production cost of treated water (€/$m^3$) |
| Billed consumption ($m^3$/year) | Losses in mains (% of real losses) | Investment in pressure control (€) |
| Billed consumption based on estimates (%) | Losses in storage water tanks (% of real losses) | Average cost of issuing a customer bill (€/bill) |
| Unauthorized consumption ($m^3$/year) | Average cost of replacing/installing a water meter (€/meter) | Average number of bills issued annually to each customer |
| Unmetered authorized consumption ($m^3$/year) | Estimated number of consumption points without a water meter | Average cost of pipe replacement (€/km) |
| Unbilled metered consumption ($m^3$/year) | Average age of customer meters (years) | Average cost of connection replacement (€/connection) |
| Length of the water network (km) | Storage volume of service water tanks ($m^3$ ) | Average cost of surface water tanks repair (€/$m^2$) |
| Total number of customers | Maximum surface of water tanks to be repaired ($m^2$) | Customer meters depreciation/life time (years) |
| Number of house connections | UARL (l/day/km pipe/metre of pressure) | Water tank repairs depreciation/ life time (years) |
| Average pressure in the network (m) | UARL (l/day/connection/metre of pressure) | Pressure control equipment depreciation/life time (years) |
| Average tariff (€/ $m^3$) | UARL (l/day/km of connection/ metre of pressure) | Pipes depreciation/life time (years) |

The user specifies basic data and interactively runs the program from a Control Panel where he/she can select a number of options and visualize the impact of those options over time. It is possible to specify a certain length of pipe replacement and see, for example, the impact on NRW, the future water balance and corresponding investment plan. It is also possible to optimize measures to be undertaken in order to maximize the absolute savings (benefit minus cost) or the relative savings (benefit/cost) over a selected period of time. This optimization can be subject to a number of restrictions, namely a maximum NRW value or maximum annual investment.

Figures 3.15 to 3.19 show screenshots of the control panel during a sample optimization process.

| Other measures | Units | Value | Max | Years |
|---|---|---|---|---|
| Pipe replacement | km | 278 | 770 | 4 |
| Connections replacement | No. | 42.443 | 60.000 | 8 |
| Tank surface repair | m2 | 19.279 | 28.000 | 4 |
| Meters (replacement) | No. | 42.815 | 153.000 | 2 |
| Local inspections | No. | 2.807 | 3.000 | 1 |
| Meters (instalation) | No. | 914 | 1.000 | 1 |
| Billing | No. | 1.000 | 1.000 | 1 |
| (Benefit-Cost)/year | | 1.339.666 € | | |
| (Benefit/Cost)/year | | 2,2 | | |

| Benefit-Invesment over | 10 | years | 7.132.247 € |
|---|---|---|---|

**Pressure + Other Measures**

| | | | |
|---|---|---|---|
| Total Investment | 17.626.573 € | | |
| Total Cost/year | 1.136.216 € | | |
| Total Benefit/year | 2.475.882 € | | |
| Total (Benefit-Cost)/year | 1.339.666 € | | |
| Pay back Investment (years) | 7,1 | | |
| Return on invest. after 1 year | -15.150.691 € | -86% | of total investment |
| Return on invest. after 2 years | -12.674.809 € | -72% | of total investment |
| Return on invest. after 3 years | -10.198.927 € | -58% | of total investment |
| Return on invest. after 4 years | -7.723.045 € | -44% | of total investment |
| Return on invest. after5 years | -5.247.163 € | -30% | of total investment |
| Return on invest. after 10 years | 7.132.247 € | 40% | of total investment |

**Performance indicators**

| | Before | After |
|---|---|---|
| UARL (m3/year) | 886.074 | 886.074 |
| CARL (m3/year) | 3.647.044 | 1.654.780 |
| ILI | 4,1 | 1,9 |
| | | |
| NRW % | 22,7% | 11,1% |
| NRW (m3/day) | 13.382 | 5.944 |
| NRW (€/year) | 3.933.590 | 1.691.138 |
| NRW (l/connection/day) | 223 | 99 |

**Figure 3.15** Control Panel of the computer program.

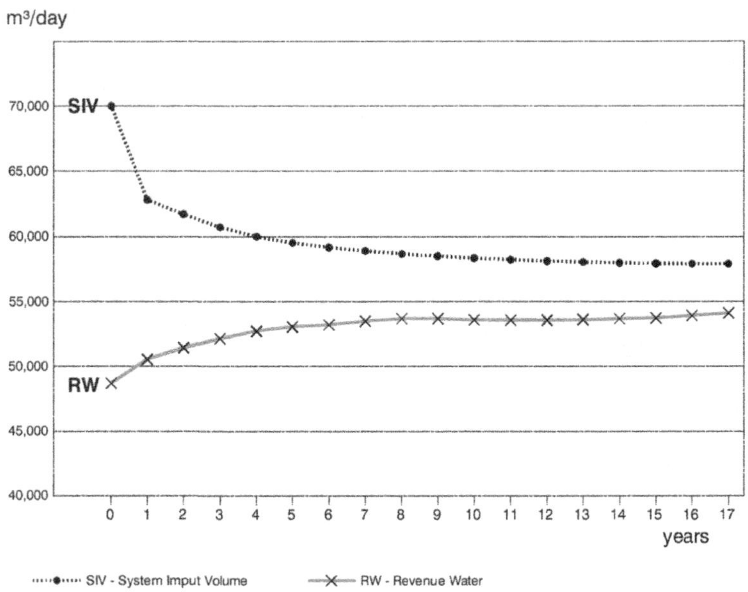


**Figure 3.16** Control Panel: Evolution of SIV and RW (m³/day).

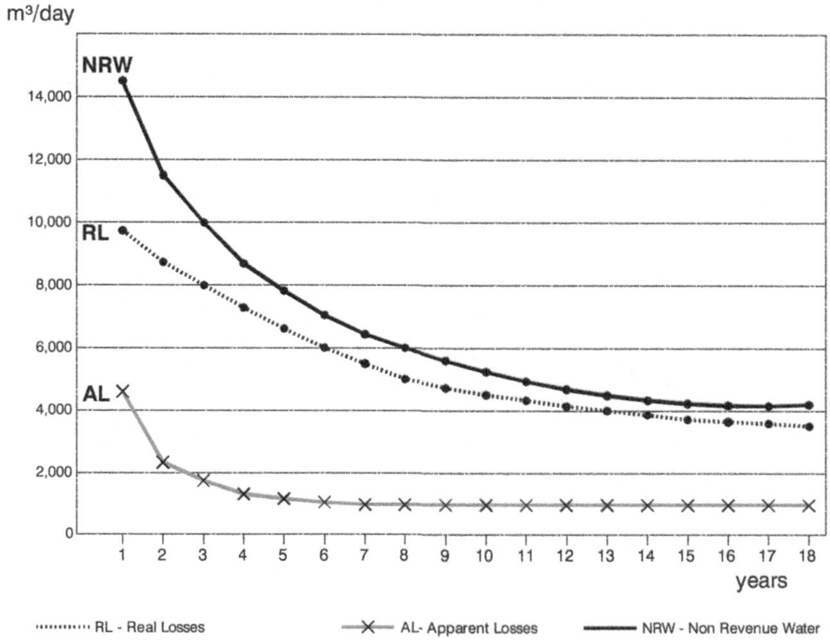


**Figure 3.17** Control Panel: evolution of Real Losses, Apparent Losses and NRW ($m^3$/day).

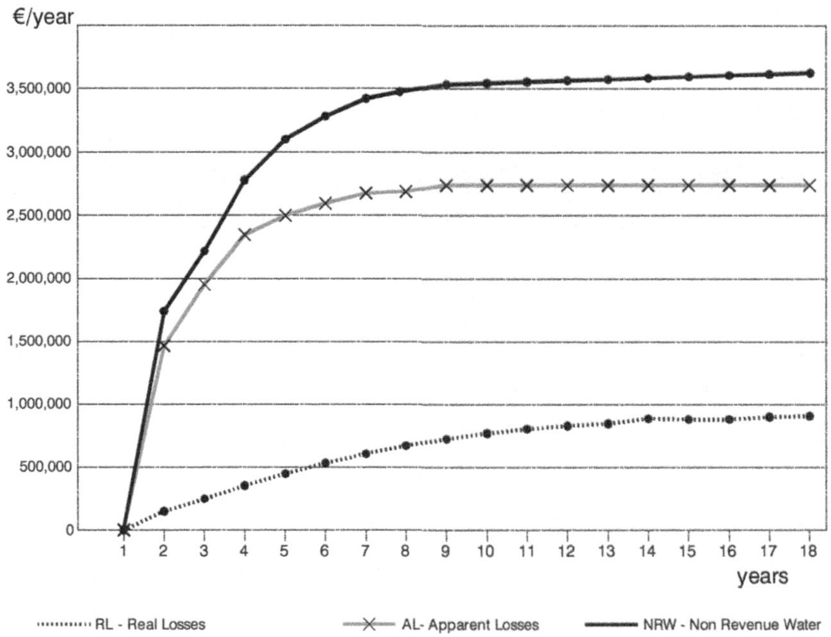


**Figure 3.18** Savings due to reduction of Real Losses, Apparent Losses and NRW (€/year).

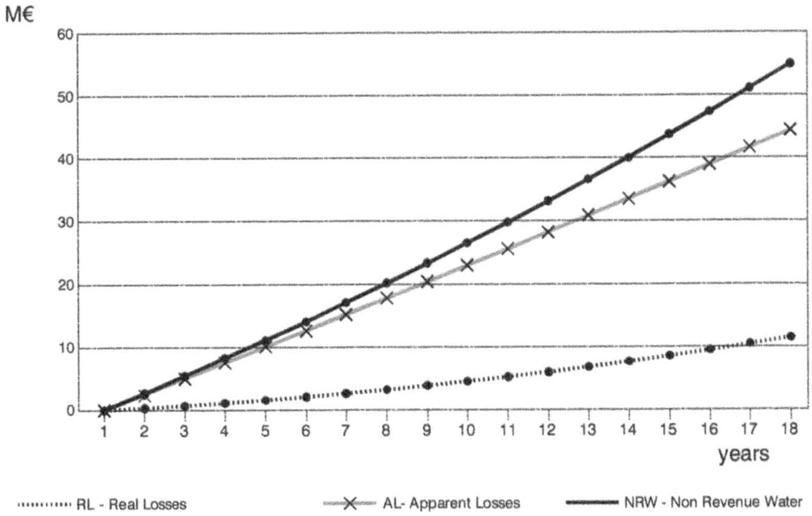


**Figure 3.19** Accumulated savings due to NRW reduction (million €).

### 3.8.3 Measures

An NRW reduction strategy is materialised as a set of measures. As previously mentioned, each of these measures has a cost that can be specified by the user as well as a corresponding benefit. The benefits follow a law of diminishing returns that can be specified by the user. Each of the NRW components requires a specific treatment, with its own agenda and commitment of resources, and can be performed through a number of specific measures (Table 3.5).

**Table 3.5** Measures to reduce NRW.

| Types of water loss | Measures 1 | Measures 2 | Measures 3 | Measures 4 |
|---|---|---|---|---|
| Leaks and bursts | Pressure control | Fast repair times | Active Leakage Control | Replacement/ repair |
| Water loss in storage tanks | Eliminating overflows | Repair of water tanks | | |
| Customer metering errors | Inspection of meters | Replacement of meters | | |
| Unauthorized consumption | Inspection of customer installations | | | |
| Consumption points without meters | Installation of meters | | | |
| Unbilled metered consumption | Billing for all consumptions | | | |

#### 3.8.3.1 Basic measures

There are certain pre-requisites for any NRW reduction strategy which are assumed as essential investments for the success of the above-mentioned specific measures, namely:

- A Central Control Room;
- Monitoring and telemetry;
- ALC - Active leakage control;
- GIS – Geographical Information System;
- Pressure management.

The central control room, monitoring and telemetry systems, ALC equipment and GIS may be considered as one-off investments which are essential to any NRW reduction process. Pressure management involves a number of activities and diversified equipment that result in the reduction of NRW, both through leakage and bursts. Pressure management systems are complex and require site-specific design by specialists in close connection to the suppliers of patented equipment and specialized software. Pressure management is known to be highly cost effective, with very low pay-back periods, therefore, for the purpose of strategy definition, pressure management can also be considered as a pre-requisite for any NRW reduction strategy. The corresponding investment costs are determined by local studies and are an input to the program.

Table 3.6 presents an example of a simple payback calculation for a pressure management investment.

**Table 3.6** Control Panel: pressure control.

| | **Pressure control** | |
|---|---|---|
| Pressure (current) | 35 | m |
| Pressure (new) | 20 | m |
| NRW reduction ($m^3$/d) | 6,698 | $m^{3/}$day |
| NRW reduction (€/year) | 977,835 | €/year |
| Investment | 1,000,000 | € |
| Cost | 133,833 | €/year |
| Benefit | 977,835 | €/year |
| Payback | 1.2 | years |

#### 3.8.3.2 Priority measures

The priority measures to be taken are mainly management measures, which are labour intensive but imply relatively low investment. These activities include a number of actions, namely:

- Repair all visible leaks;
- Elimination of overflows in water tanks;
- Universal billing/metering;
- Local inspection of meters;
- Installation of new meters and/or data-loggers at the major customers;
- Identification of anomalous consumption patterns.

#### 3.8.3.3 *Measures with specified cost/benefit relations*

There are seven major activities for which it was possible to establish cost/benefit relations in this program, in terms of NRW reduction (Table 3.7). For each of these measures there are a maximum number of units that can be accomplished: total and annually. The total limit corresponds, for instance, to the length of the existing pipe network, whilst the annual limits are specified by the user depending on the physical capacity of execution or on the perceived impact on city life (e.g. pipe replacement/rehabilitation). It is possible to achieve a long term ELL taking into consideration all investment and operation costs for, say, 100 years, the optimistic useful life of pipes and other long lasting assets. However, over such a long period of time, it is likely that there will be substantial alterations to the current conditions. For instance, over a period of 100 years it is possible, for example, that the institutional model will change or that there is a merger with neighbouring systems.

**Table 3.7** Control Panel: medium and long term measures.

| Measures | Units | Value | Max | Time (years) |
|---|---|---|---|---|
| Pipe replacement | km | 278 | 700 | 4 |
| Connections replacement | No. | 42,699 | 60,000 | 8 |
| Tank surface repair | $m^2$ | 18,314 | 28,000 | 4 |
| Meters (replacement) | No. | 43,302 | 153,000 | 2 |
| Local inspections | No. | 2,807 | 3,000 | 1 |
| Meters (installation) | No. | 914 | 1,000 | 1 |
| Billing | No. | 1,000 | 1,000 | 1 |
| (Benefit-Cost)/year | €/year | € 1,399,773 | N.A. | |
| Investment | M€/year | 17.7 | N.A. | |
| Pay back | years | 7.1 | N.A. | |

Investment decisions regarding NRW reduction based on uncertain projections over such a long period of time are difficult to support if they are substantially different of those corresponding to, say, a 20 - 30 year period. In any case, the decision is upon the user and if there are unforeseen alterations, the corresponding investment and operational costs will hopefully be covered by future tariffs.

### 3.8.4 Results

Essentially, owners and managers want to know what to do to reduce NRW, how far it is physically and economically feasible to go, how much it will be necessary to spend each year in each activity, and what results could be expected. The main novel aspect of the proposed computer program is the ability to define measures for NRW reduction and the corresponding targets as a result of a formal optimization algorithm. Unlike conventional approaches, it is not necessary to estimate or anticipate the target, namely the ELL, which is just one of the possible targets and depends very much on the options assumed by the user in their calculations.

The results presented below should be analyzed on a comparative basis, as they depend on the previous definition of many variables which are site specific and depend on the preferences of the top manager using the program. The most interesting feature is, perhaps, its flexibility and ability to practice, that is, the capability of evaluating alternative realistic targets, using different variables and alternative objective functions and restrictions in the optimization algorithm.

Tables 3.8 and 3.9 show initial and final values of key performance indicators and the corresponding investment, costs and benefits as a result of the optimization algorithm.

**Table 3.8** Results: performance indicators.

| Indicators | Before | After |
|---|---|---|
| UARL ($m^3$/year) | 869,978 | 497,130 |
| CARL ($m^3$/year) | 6,004,250 | 1,258,833 |
| ILI | 6.9 | 2.5 |
| NRW % | 30.0% | 7.5% |
| NRW ($m^3$/day) | 21,000 | 4,247 |
| NRW (€/year) | 5,723,200 | 1,085,984 |
| NRW (l/connection/day) | 350 | 71 |

**Table 3.9** Results: investment, costs and benefits.

| | Pressure control + other measures |
|---|---|
| Total Investment | € 17,365,182 |
| Total Cost/year | € 1,188,433 |
| Total Benefit/year | € 2,476,515 |
| Total (Benefit-Cost)/year | € 1,288,082 |
| Payback on investment (years) | 7.0 |

Figure 3.20 shows the evolution of SIV and RW over time for the optimized solution, leading to the ELL, whilst Figure 3.21 shows the annual investment required to implement each of the measures for the optimized solution.

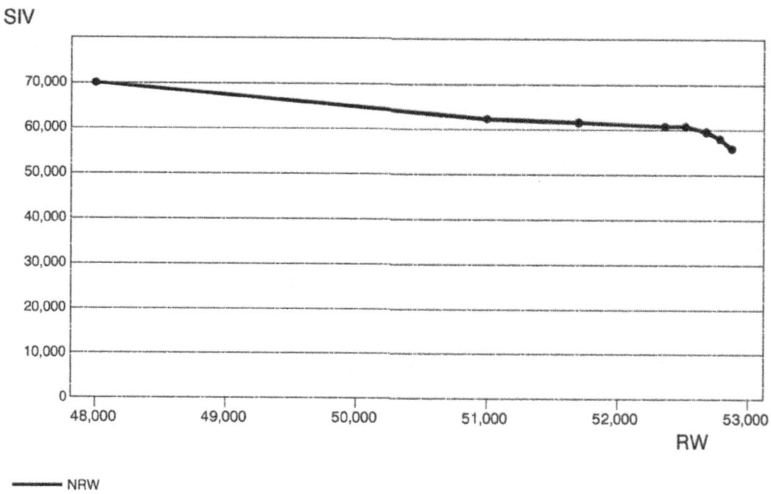


**Figure 3.20** Evolution of the SIV and RW over time leading to the ELL.

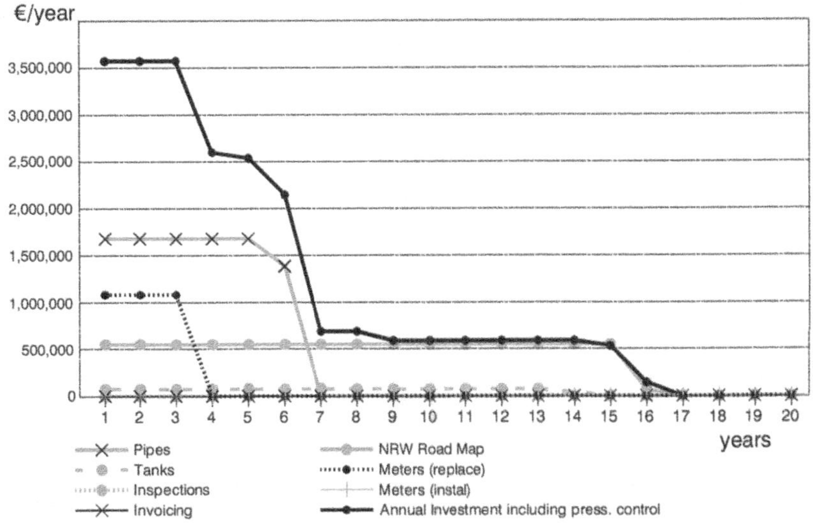


**Figure 3.21** Annual investment required to implement each measure (€/year).

The program makes it very easy to perform a sensitivity analysis for any of the variables that determine the costs and benefits. Table 3.10 shows the impact of the production cost of the water entering the system (€0.1, €0.4 and €0.6/m$^3$) on the optimized final values of the NRW (immediately after the implementation of the measures), the ILI and the percentage of network pipes replaced. As expected, for example, the higher the price of the treated water, the more interesting replacing pipes becomes; this results in lower values of NRW but with significantly higher investment and not leading to better financial results, namely in terms of payback.

**Table 3.10** Sensitivity analysis on the price of the treated water.

| | **Raw water at €0.10/m$^3$** | **Raw water at €0.40/m$^3$** | **Raw water at €0.60/m$^3$** |
|---|---|---|---|
| Investment (M€) | 3.1 | 17.1 | 23.4 |
| Final NRW (%) | 18.6% | 11.0% | 9.1% |
| Final NRW (m$^3$/d) | 10,882 | 5,880 | 4,741 |
| ILI | 3.9 | 1.8 | 1.4 |
| Pipe replacement (km) | 1 | 278 | 442 |
| Pipe replacement (% of network) | (0.0%) | (10.1%) | (56.8%) |
| Benefit-cost/year | €875,311 | €1,205,638 | €1,658,034 |
| Payback on investment (years) | 3.5 | 14.2 | 14.1 |

Figures 3.22, 3.23 and 3.24 show the final water balances corresponding to the same values of the water entering the system.

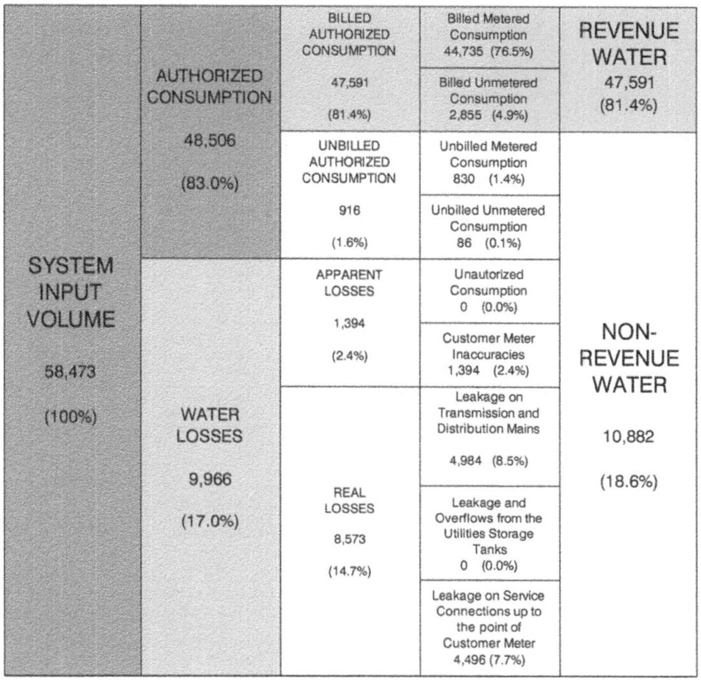


**Figure 3.22** Sensitivity analysis: water balance for € 0.1/m³.

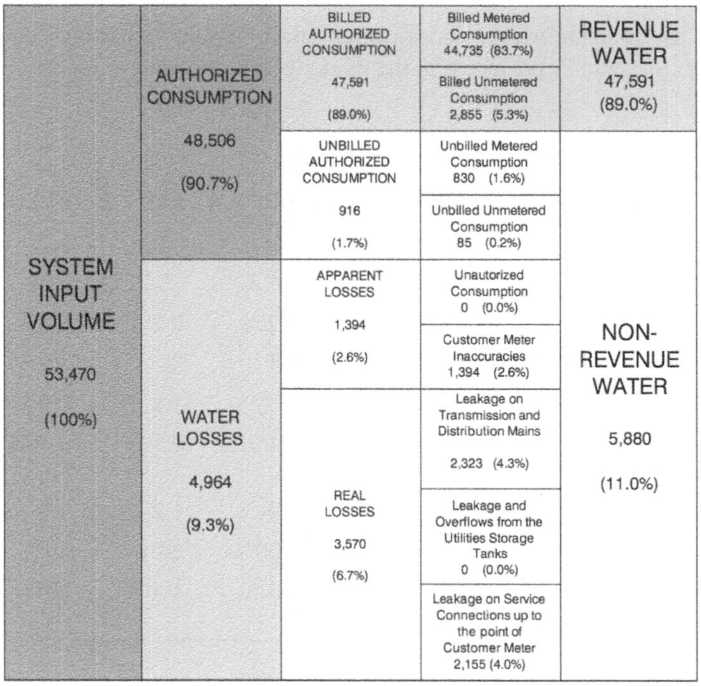


**Figure 3.23** Sensitivity analysis: water balance for € 0.4/m³.

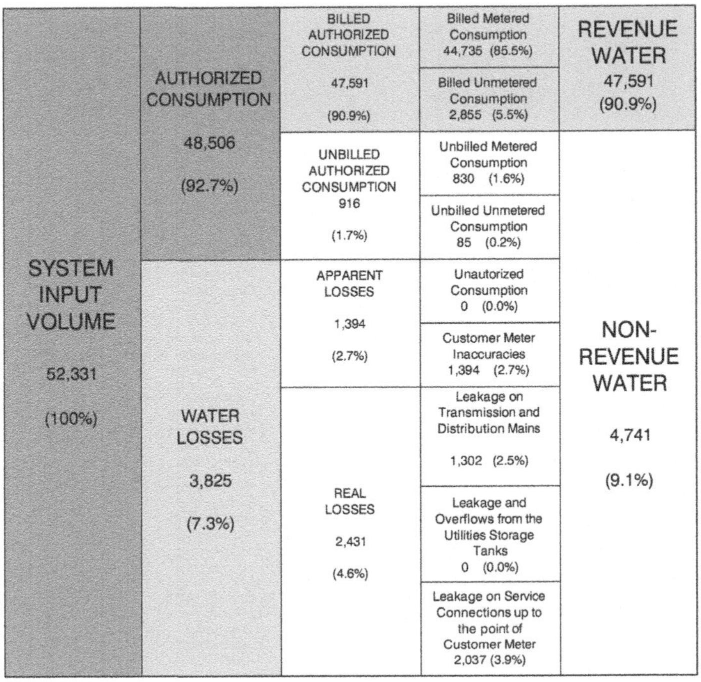


**Figure 3.24** Sensitivity analysis: water balance for € 0.6/m$^3$.

Table 3.11 shows the comparative results of running the program, departing from an initial situation inspired by one of the case studies presented in the book, in the following sequence:

**Table 3.11** Target setting alternatives (1).

| | Initial Situation | ELL (p= 35m) | ELL + A | ELL + B | ELL + C | ELL (p= 25m) |
|---|---|---|---|---|---|---|
| Pressure (m) | 35 | 35 | 35 | 35 | 25 | 25 |
| Total Investment (M€) | N.A. | 17.1 | 34.3 | 23.7 | 17.6 | 10.0 |
| NRW (%) | 22.7% | 11.0% | 7.8% | 9.2% | 8.8% | 11.2% |
| NRW (l/conn/day) | 223 | 98 | 67 | 82 | 77 | 100 |
| ILI | 5.8 | 1.8 | 1.1 | 1.8 | 2.6 | 3.7 |
| Annualized Benefit-Cost (M€/year) | N.A. | 1.205 | 0.937 | 0.521 | 0.906 | 1.031 |
| Ratio Benefit/Cost | N.A. | 2.1 | 1.6 | 1.2 | 1.9 | 2.3 |
| Pipe replacement | N.A. | 36% | 100% | 36% | 36% | 11% |
| Customers' Meters Replacement | N.A. | 28% | 28% | 100% | 28% | 28% |

Notes:
A - Further replacement of the remaining pipes
B - Further replacement of the remaining customers' meters
C - Further reduction of average pressure to 25 m

1. Calculating the ELL for an average pressure of 35m;
2. (1) + further replacement of the remaining pipes;
3. (1) + further replacement of the remaining customers' meters;
4. (1) + further reduction of pressure from 35 m to 25 m;
5. Calculating ELL for an average pressure of 25 m.

The results show significant differences in terms of the total investment required, on the NRW and on the cost/benefit relations. The absolute values, as mentioned below, are not important for the purposes of this evaluation, as they depend on a number of site specific variables that have been selected. It is particularly interesting to note the importance of pressure reduction: the earlier it is accomplished, the higher the cost effectiveness and the lower the investment.

Table 3.12 shows the comparative results of running the program for the following situations:

1. Calculating the ELL for an average pressure of 35m;
2. Optimizing for a maximum total investment of M€ 10;
3. Optimizing results in 4 years' time;
4. Optimizing for NRW = 15%.

**Table 3.12** Target setting alternatives (2).

| | ELL | Optimizing for total investment = M€10 | Optimizing results in 4 years | Optimizing for NRW = 15% | Optimizing for ILI = 2 |
|---|---|---|---|---|---|
| Pipe replacement | 36.2% | 11.6% | 0.1% | 7.9% | 31.5% |
| Connections replacement | 70.7% | 61.5% | 0.0% | 57.0% | 69.0% |
| Tank surface repair | 69.2% | 58.5% | 0.0% | 50.5% | 63.4% |
| Meters (replacement) | 28.1% | 24.2% | 13.8% | 23.6% | 28.3% |
| Local inspections | 93.6% | 93.2% | 92.3% | 93.1% | 93.6% |
| Installation of new meters | 91.4% | 90.9% | 89.6% | 90.7% | 91.4% |
| Billing new customers | 100.0% | 100.0% | 100.0% | 100.0% | 100.0% |
| Total investment (M€) | 17.6 | 10.0 | 1.8 | 8.6 | 16.3 |
| NRW (%) | 11.1% | 14.3% | 20.0% | 15.0% | 11.6% |
| NRW (l/conn./d) | 99 | 132 | 197 | 140 | 104 |
| ILI | 1.9 | 2.6 | 4.1 | 2.6 | 2.0 |
| (Benefit-cost)/year (M€/year) | 1,340 | 1,294 | 1,006 | 1,275 | 1,338 |
| Benefit/cost | 2.2 | 2.6 | 3.7 | 2.6 | 2.2 |
| Payback (years) | 7.1 | 4.7 | 1.3 | 4.2 | 6.7 |

This example shows that there are many alternatives to NRW target setting. ELL is just one of the possibilities and, even so, its value depends very much on the values of the site-specific and non-site specific variables that may be used in its calculation. In fact, there are many different reasons and motivations for top management setting NRW reduction targets.

The program's role is just to assist in this process, giving the answers to the what-if questions. As expected, the results can not be better than the top manager using it. In its current version, the program must be used by a NRW consultant trained in the use of the program, working remotely in conjunction with top management.

From personal experience, it is not important to develop the program to a full interactive version, as if it were a computer game, to be used directly by top managers. This would require an extremely time consuming effort to manage all the possible exceptions and avoiding the introduction of unrealistic values for the variables.

The program could help to save very large amounts of money, and this is certainly compatible with a delay of hours or days between iterations; this period is necessary so that the top manager's team can produce new data sets based on the previous results. The results of the program may be used as inputs to GIS based procedures to select, for example, the pipes or customer meters to replace, or to identify the suspicious installations where potentially situations of unauthorized consumption occur; this is only possible through the use of the specific water utility customer and infrastructure databases.

## 3.9 REFERENCES

Alegre H. and Almeida M. C. (2009). Strategic Asset Management of Urban Water Infrastructures. IWA Publishing, London, UK.

Alegre H., Hirner W., Baptista J. M. and Parena R. (2000). Performance Indicators for Water Supply Services. IWA Publishing, London, UK.

Brothers K. (2003). A practical approach to water loss reduction. IWA Water Loss Task Force, *Water 21* Magazine, Special Series Water Loss, Article No.1, Issue 5.3, June 2003, pp 54-55.

CNQA National Quality Committee ABES (2013). Guide PNQS 2013, ed. ABES, Rio de Janeiro, Brazil.

Fanner P. (2004). Assessing Real Losses: a Practical Approach. IWA Water Loss Task Force, *Water 21* Magazine, Special Series Water Loss, Issue 6.2, April 2004, pp 49-50.

Farley M. (2001). Leakage Management and Control – A Best Practice Training Manual. World Health Organization, Geneva, Switzerland.

Farley M. (2010). Are there alternatives to DMA? *Asian Water*, Vol. December 2010, pp. 10-16.

Farley M. and Trow S. (2003). Losses in Water Distribution Networks – A Practitioner's Guide to Assessment, Monitoring and Control. IWA Publishing, London, UK.

Farley M., Wyeth G., Ghazali Z. B., Istandar A. and Singh S. (2008). The Manager's Non-Revenue Water Handbook – A Guide to Understanding Water Losses. Ranhill Utilities Berhad and United States Agency for International Development.

Frauendorfer R. and Liemberger R. (2010). The Issues and Challenges of Reducing Non-Revenue Water. Asian Development Bank, Mandaluyong City, Philippines.

Hamilton S. and Charambolous B. (2013). Leak Detection: Technology and Implementation. IWA Publishing, London, UK.

IWA Water Loss Task Force (2008). Target Setting Initiative Consultation Paper, Bucharest.

IWA Water Management Specialist Group (2013). IWA Water Utilities Assessment Matrix: Quick Reference Guide. www.iwahq.org, http://www.iwahq.org/7m/networks/specialist-groups/list-of-groups/urban-water-systems.html, last accessed on 30/5/2014.

Kingdom B., Liemberger R. and Marin P. (2006). The Challenge of Reducing Non-Revenue Water in Developing Countries – How the Private Sector Can Help: A Look at Performance-Based Service Contracting, Water Supply and Sanitation Sector Board Discussion Paper Series, Paper No. 8, World Bank, Washington, D.C., USA.

Lambert A. (2001). What Do We Know About Pressure: Leakage Relationships? Proceedings of the IWA Conference on Systems Approach to Leakage Control and Water Distribution Systems Management, Brno.

Lambert A. (2003). Assessing non revenue water and its components: a practical approach. *Water 21* Magazine, IWA, Issue 5.4, August 2003, pp. 50-51.

Lambert A., Brown T. and Weimer D. (1999). A review of performance indicators for real losses from water supply systems. *Journal of Water Supply: Research and Technology – Aqua*, **48**(6), 227-237.

Lambert A., Charambolous B., Fantozzi M., Rizzo A. and Galea St John G. (2014). 14 Years Experience of Using Best Practice Water Balance and Water Loss Performance Indicators in Europe. http://www.leakssuite.com/outreach/free-papers-2014/, last accessed on 30/5/2014.

Lambert A. and Hirner W. (2000). Losses from Water Supply Systems: Standard Terminology and Recommended Performance Measures. IWSA Blue Pages, London, UK.

Lambert A. and Liemberger R. (2005). Accuracy Limitation of the ILI – Is It an Appropriate Indicator in Developing Countries? IWA Water Loss Conference Halifax, World Bank.

Lambert A. and McKenzie R. (2002). Practical Experience in Using the Infrastructure Leakage Index, IWA Leakage Management Conference: A Practical Approach, Cyprus, November 2002.

Lambert A. and Morrison J. (1996). Recent developments in application of bursts and background estimations (BABE) concepts for leakage management. *CIWEM Journal*, **10**(2), 100-104.

Liemberger R. (2006). Introduction to Water Loss Analysis and Reduction, Water Loss Task Force, IWA Specialist Group for Efficient Operation and Management of Urban Water Supply Systems and World Bank Institute. Water Loss Workshop at the IWA World Water Congress, Beijing, India, September 2006.

Liemberger R. (2007). Water Loss Performance Indicators, IWA Specialised Conference Water Loss 2007, Bucharest, September 2007.

Marin P. (2009). Public–Private Partnerships for Urban Water Utilities: A Review of Experiences in Developing Countries, The World Bank and PPIAF Public–Private Infrastructure Advisory Facility, Washington, D.C., February 2009.

McKenzie R. and Wegelin W. (2002). Leakage Reduction Trough Pressure Management in South Africa, Proceedings of IWA Conference Leakage Management: A Practical Approach, Cyprus.

Morrison J. (2004). Managing leakage by district metered areas: a practical approach. IWA Water Loss Task Force, *Water 21* Magazine, Special Series Water Loss, Article No. 5, Issue 6.1, February 2004, pp.44-46.

Morrison J., Tooms S. and Rogers D. (2002). DMA Management Guidance Notes, DMA Team of the IWA Water Loss Task Force, Speciality Conference, Cyprus.

Parker J. (2009). Collecting information to optimize a mains replacement strategy. *Water Asset Management International*, **5**(3), 18-20.

Pearson D. and Trow S. (2005). Calculating Economic Level of Leakage, Proceedings of IWA Water Loss Task Force Leakage 2005 Conference, Halifax.

Pilcher R., Hamilton S., Chapman H., Field D., Ristovski B. and Stapely S. (2007). Leak Location and Repair Guidance Notes, IWA Water Loss Task Force, www.sswm.info/library/3122, last accessed on 2014.05.30.

Rizzo A., Pearson D., Stephenson M. and Harper N. (2004). Apparent water loss control: a practical approach. IWA Water Loss Task Force, *Water 21* Magazine, Special Series Water Loss, Article No.7, Issue 6.3, June 2004, pp. 44-45.

Savic D., Lambert A. and Kapelan Z. (2005). Water losses management and leakage detection techniques for water distribution systems. *Water & Sewerage Journal*, Vol. 2, pp. 25-27.

Thornton J. (2003). Managing leakage by managing pressure. *Water 21* Magazine, Issue 5.5, October 2003, pp. 43-44.

Thornton J. (2008). Water Loss Control, 2nd edn. McGraw-Hill, USA.

Thornton J. and Lambert A. (2005). Progress in Practical Predictions of Pressure: Leakage, Burst Frequency and Consumption, IWA Water Loss Leakage 2005 Conference Proceedings, Halifax, Nova Scotia, September 2005.

Thornton J. and Lambert A. (2006). Managing pressure to reduce new breaks. *Water 21*, Magazine, Issue 8.6, December 2006, pp. 24-26.

Tripartite Group (2002). Best Practice Principles in the Economic Level of Leakage Calculation, DEFRA – Department for Environment, Food and Rural Affairs, Environmental Agency and Ofwat – Office of Water Services, March 2002.

Trow S. and Pearson D. (2010). Setting Targets for Non Revenue Water Reduction, *Water 21* Magazine, Issue 12.2, April 2010, pp. 21-22.

World Bank Institute (2005). NRW Training Module 6: Performance Indicators, pro-wat.com/uoloads/media/prowatwl book_english.pdf, last accessed on 2014.05.30.

Wu Z., Farley M., Turtle D., Kapelan Z., Boxall J., Mounce S., Dahasahasra S., Mulay M. and Kleiner Y. (2011). Water Loss Reduction. Bentley Institute Press, Exton, Pennsylvania.

# Chapter 4

# Restructuring Water Utilities

## 4.1 INTRODUCTION

Restructuring, turning around management or fundamental change to a water utility is always a very complex process. There are invariably many specific technical, economic, social, environmental and political problems to resolve and complex decisions to make involving local, regional and central levels of government. Moreover, some of these decisions are related to highly sensitive subjects such as public/private interfaces, poverty related issues and access to water services as a human right.

This chapter addresses a number of key issues related to the complex task of restructuring water companies using the theoretical fundamentals of change management, accumulated knowledge of the water industry, personal experience in a number of water utilities and, most importantly, the lessons that can be extracted from the case studies presented. Some of the issues analysed are: privatization, political options and economies of scale, the water systems of the future and the integrated management of the urban water cycle.

The chapter also presents specific recommendations related to a number of features that are important to the modernization of old organizations including reaching profitability, dealing with overstaffing, designing and supervising new works, increasing water supply system reliability whilst introducing innovation and controlling performance. It also analyses the essential practical aspects of relations with stakeholders, namely customers, owners, regulator, workers and their unions, suppliers, the media and Environmental Non-Governmental Organizations (ENGOs).

Finally, a simplified restructuring model for water utilities is outlined and tested in real case studies. The model is based on the assumption that a four to six year transition period to reach sustainability and excellence is necessary. Acceptance of the indispensability of this transition period to implement water utility restructuring is the first condition for success. Management of the transition or restructuring period is based on four strategic complementary actions; (1) increasing efficiency of the water utility, mainly through the reduction of NRW, energy and labour; (2) setting sustainability limits for the water utility (always a difficult political decision); (3) investing in the expansion and rehabilitation of the system; and, finally, (4) adjusting tariffs to ensure full cost recovery. These apparently simple and common sense measures specific to the water utilities sector are very complex and require political support, leadership, people, financial resources, time and a favourable context.

Integration of the process follows change management theory, as outlined for example in the eight points of John Kotter's methodology. Basically, before a commitment is made to restructure poorly performing water utilities, there are a number of critical questions to ask:

- Is it possible to have a 4 to 6 year transition period with management stability to implement the restructuring process?
- Is it possible to redefine the sustainability boundaries of the systems, if they are too small?
- Is it possible to obtain external support through transfers or taxes in cases of extended poverty and/or a very dispersed rural or isolated population?
- Is there a consensus on utility targets and future tariffs?

If the answer to any of the above questions is no, noting that some are political in nature, the chances of the success of the restructuring process may be slimmer.

However, politics is an art, the art of the possible, and it is often unavoidable to accept compromise solutions and explore less ambitious windows of opportunity. In particular, it may be politically very difficult to ensure that there will be a sufficiently long stability period. Alternatively, definition of the future boundaries of the water utilities may be predefined through political negotiation at an early stage, but this is usually very difficult to achieve, as it touches vital institutional interests. From personal experience, the leader of any restructuring may assist in reaching a political consensus on this issue, proposing technically sound win-win solutions, for all the parties involved, at an early stage.

## 4.2 SPECIAL RESTRUCTURING ISSUES

### 4.2.1 The great complexity of the water utilities sector

For many years, I have wondered why the local beer is almost always good and reliable anywhere in the world and why the same cannot be said about the local tap water whilst, apparently, tap water is much simpler to produce than beer.

Indeed, potable water is a product with approximately the same specifications all over the world; it is essential to life and there is no substitute for it. There are well proven technologies to transform virtually any kind of raw water, more or less polluted, into potable water; therefore the problem is definitely not technological. Pipe networks are underground and the service is run by water utilities. Therefore, the reasons behind the current situation of poor quality water services in many parts of the world should be related to problems in the networks, in terms of construction and maintenance as well as water utilities management, certainly on side B of the vertical line in Figure 4.1, which represents in a simplified way the major components and inputs of a water supply system.

These problems are the result of decisions taken at local, regional and national level, have technical, economic, environmental, social, political and emotional dimensions, and are strongly affected by poverty related issues. Moreover, water utilities can be owned and managed by public or private entities, under many different models, and are frequently strongly affected by political cycles (Figure 4.2).

In the past, one of the major tasks for the management of a water utility was maintaining an equilibrium at all times between the legitimate but sometimes conflicting interest of the owners, customers and the workers. This role has become significantly more difficult with the involvement of a new player, namely the regulator, which limits the options available to the owners and management (Figure 4.3). This high level of complexity explains why some problems appear not to have a timely solution and why the majority of the restructuring processes, which, from personal experience, should take up to six years, may last indeed much longer or possibly prove not to be successful at all.

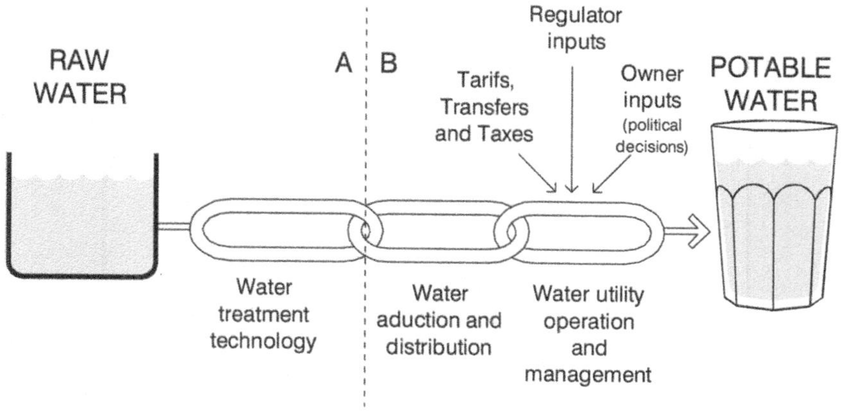


**Figure 4.1** Complexity of producing potable water.

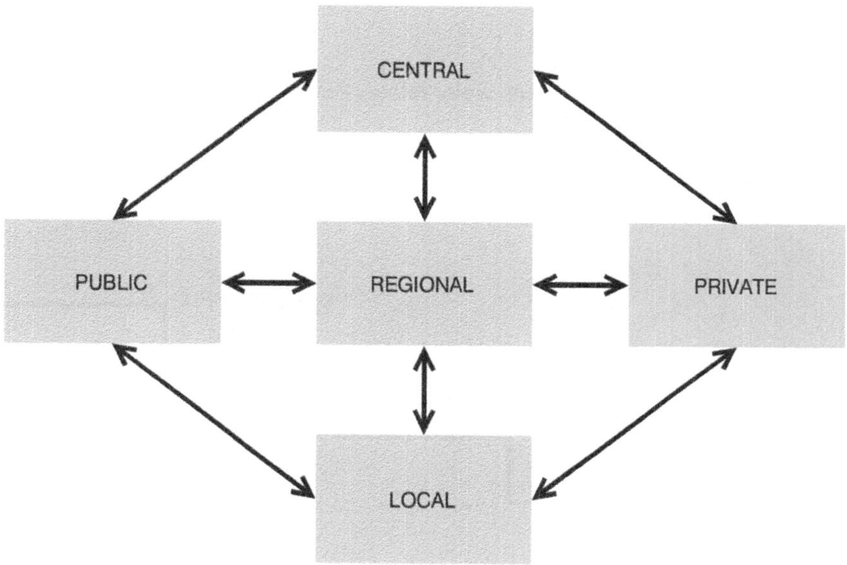


**Figure 4.2** Complex interfaces interfering with water related decisions (1).

The word water is used to designate very different things, namely raw water (which is a natural resource), bottled water (which can be purchased in shops) and tap water (which, like bottled water, is an industrial product and costs money to produce and deliver to customers). There is confusion between the concepts of access to water and access to water services, and water related decisions involve technical, economic, social, environmental and political components (Figure 4.4).

Included in the political component in particular are opposing ideological positions regarding privatization and free access to water services. This complexity leads to emotional attitudes in the population, amplified or induced by the media and interfering with rational decisions, for example when it comes to privatizing water utility services or shutting off service to non-paying customers.

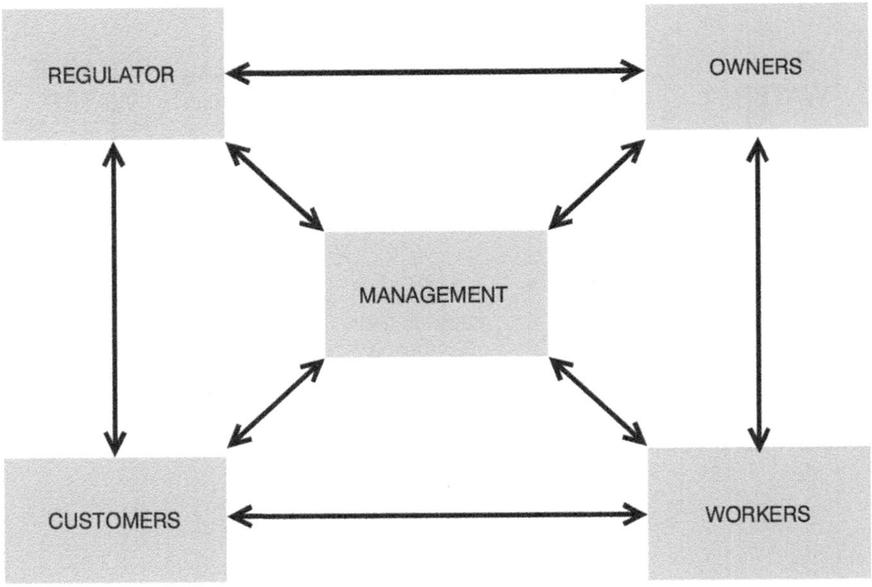


**Figure 4.3** Complex interfaces interfering with water related decisions (2).

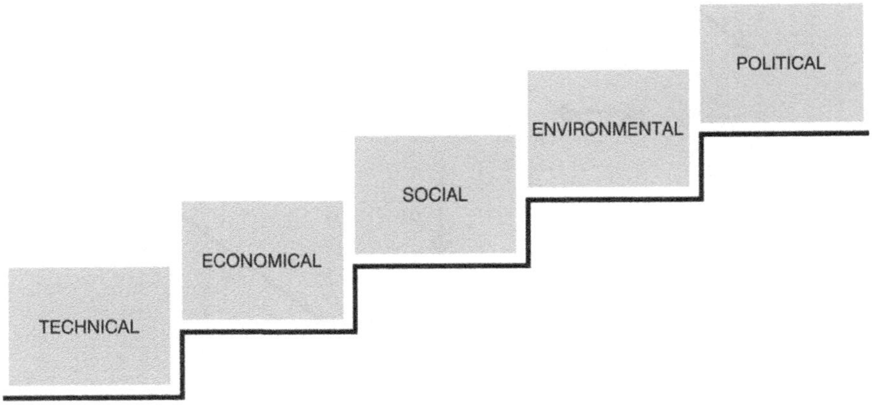


**Figure 4.4** Levels of decision making on water issues.

No matter how difficult it may be to restructure a single water utility, it is important to realize that the water sector consists of many separate water utilities and some decisions which are made concerning one of them may interfere with others. In particular, in a restructuring process at the national level, it would certainly be easier to deal separately with the large water utilities first, but then the smaller and more problematic ones would be left behind, becoming worse and worse, eventually leading to unacceptable public health, social and environmental problems. In such cases, to avoid service deterioration, it would be necessary for the small municipalities to permanently subsidise their unsustainable water utilities, to take the difficult decision of merging with neighbour larger ones, or transferring responsibilities to the state, if this possibility exists.

Alternatively, if there is a political decision to restructure the water sector of an entire country or large region, there will be two major issues to deal with, namely: the future role of the municipalities, which usually have the ultimate political responsibility for water services; and the fact that the larger municipalities most likely have already found sustainable solutions for their water utilities and may not be willing to go through a restructuring process in which they have to deliver mostly one-way solidarity to others.

One thing appears to be clear: in developed countries the restructuring phase has often been undertaken many years ago and currently, in general, there are relatively large sustainable systems in operation in urban areas, publicly (most of them) or privately owned and managed, and where service quality and tariffs are no longer problematic issues. In many developing countries, however, universal coverage and access to water services is still a major problem, and there are typically many large and small unsustainable systems calling for restructuring.

However, more than half of the world's population lives in urban areas and future demographic growth is expected to be concentrated in these urban areas, mainly in the slums of the megacities in developing countries. A recent book, coordinated by Veiga da Cunha, addresses in depth the major issues related to water and water security (Gulbenkian Think Tank on Water and the Future of the Humanity, 2014).

Climate variability and change have important consequences on water availability and variability at four different timescales (van den Hurk and Jacob, 2009), all interfering with water utility management: (1) synoptic (droughts, floods and other individual events); (2) seasonal (consistently higher or lower precipitation); (3) pluriannual or decadal (higher or lower than average precipitation); and (4) multi-decadal or century (affecting storage, water supply and flood protection). Dry areas will tend to get dryer whilst already wet areas may get wetter; weather patterns may also shift with climate change, making predictions more difficult.

Rainfall variability may induce even greater variability of river flows. In semi-arid regions, even a small reduction in rainfall may result in significant flow reductions and can cause rivers to dry up completely. On the other hand, in colder climates, the snowmelt period may change and affect seasonal flows which are normally used of fill reservoirs for water supply and other water uses. The frequency and magnitude of droughts and floods is likely to increase, negatively affecting the existing water quantity and quality problems, namely the effectiveness and operation of the water treatment plants. Considering that a significant proportion of consumption relies on stored water, either in the ground (groundwater) or in reservoirs, the greater rainfall and flow variations may lead to the aggravation of existing water conflicts and to the creation of new ones.

The cumulative impact of demography, poverty and the adaptation to climate change will put additional pressure on the management of many water utilities, especially in the developing world (World Bank, 2002; World Bank and International Bank for Reconstruction and Development, 2008). Globalization and the current economic crisis also affect developed countries, creating new forms of poverty and leading to affordability problems that could not have been anticipated just a few years ago. In any case, more poor families will be unable to afford the full cost of water services, and it will be necessary to subsidize access and even consumption. States, donors, civil society and water utilities must cooperate to create imaginative solutions for those who are excluded: all households in all settlements, either formal or informal, should have access water services (United Nations, 2008; Albuquerque, 2012).

### 4.2.2 Privatization

Privatization, used here as a shorthand for private sector participation in the provision of water services, is one of the ways to undertake restructuring of a water utility, passing the problem to a private specialized firm instead of conducting the process under public management. The same basic change management principles apply to both public and private management but in the latter case there is usually a more clear definition of the objectives whilst there is a contract that ensures stability for a sufficiently long period of time to restructure the company, certainly longer than a political cycle.

Privatization is still a controversial and ideologically sensitive issue, with a history characterized by long and inconclusive discussions on potential advantages and disadvantages. However, independent studies comparing public and private utilities show little difference in performance between them: there are many good and bad examples of both models.

Unfortunately, early views of reputable international organizations such as the World Bank that privatization would solve the problems of universal access to water services for the poor in developing countries did not materialize. In fact, it is not realistic, perhaps, to assume that the political risks associated to poverty, war, corruption and generalized non-payment of tariffs, may be accommodated in long term public-private partnerships (PPPs). Privatization in water utilities may assume different forms and various degrees of intensity, as seen in Table 4.1. There are many variations of these models, for example the BOT (Build, Operate and Transfer) the BOOT (Build, Own, Operate and Transfer) and different forms of participation in the capital of public utilities as shareholders (with or without management responsibilities).

**Table 4.1** Some models of private participation in the water sector (Source: adapted from World Bank, 2002).

| | **Property** | **Operation** | **Investment** | **Risk** | **Duration (years)** |
|---|---|---|---|---|---|
| Outsourcing contract | Public | Shared | Public | Public | 1-3 |
| Management contract | Public | Private | Public | Public | 3-5 |
| Renting contract | Public | Private | Shared | Public | 8-15 |
| Concession | Public | Private | Private | Private | 20-30 |
| Privatization | Private | Private | Private | Private | Unlimited |

In the past there have been a number of private/public cycles in water utility management. During the early 1800s in Europe and the United States, private companies were the main water suppliers in larger cities. One hundred years later, circa 1900, the situation had changed and hundreds of private water utilities returned to public management and many other public utilities had been created. In the 1980s, a new privatization movement started again. In Britain, there was a privatization of the whole water utilities sector, including infrastructure, led by the then prime-minister Margaret Thatcher; it included the creation of a new economic regulatory agency, OFWAT– the Office of Water Services. This process and its subsequent evolution is still a case study at international level.

In the 1990s, concessions were the main model for private investment in the water sector. Enthusiasm for this model seems to have cooled from 2000 (Figure 4.5) and the most important private companies moved away from concessions in developing countries. Now, they look to be betting on other models, such as leasing, management and performance based contracts, without financial risk or the responsibility for investment (de Mästle et al., 2008; Marin, 2009; Marin et al. 2009).

Despite this constant change of hands over time, public authorities continue to dominate the ownership and management of the water utilities around the world, accounting for more than 90% of water supplied. As has happened in the past, publicly owned water utilities are signing contracts outside their boundaries where they operate as private partners.

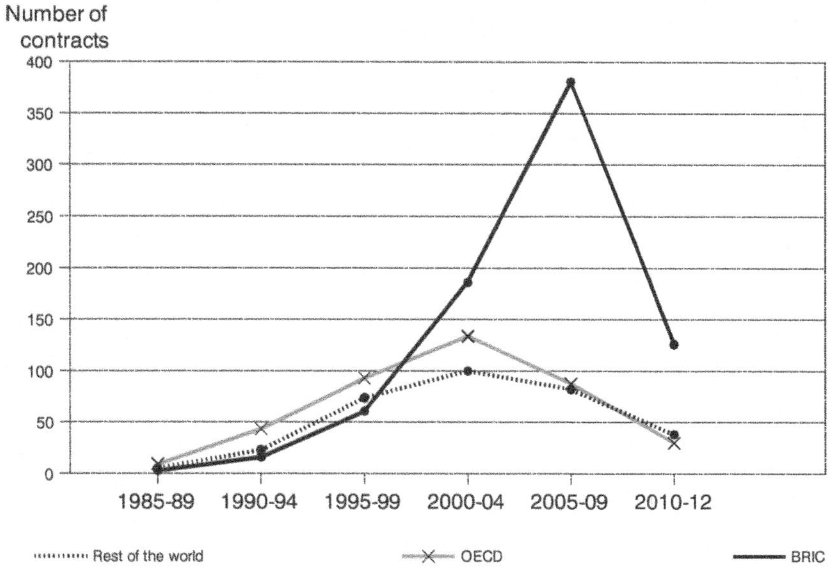


**Figure 4.5** Evolution of the number of concession contracts (Source: Pinsent Masons, 2014).

### 4.2.3 The decision of the Mayor about his/her water utility

Water utilities are, in most cases, public and municipal and, therefore, mayors, at some stage, are confronted with a difficult decision: what to do about them? The answer is difficult, because there are always a number of alternatives for each municipality. The optimal solution is, ultimately, the one that the mayor is more comfortable with and it is ultimately, a political decision. The outcome depends on the specific requirements of the municipality and also, to a great extent, on the views of the mayor: the technical, economic and legal components, although complex, are easier to deal with.

Municipalities with a population over, say, a hundred thousand people, with a high coverage of water services and many paying customers, are compatible with virtually any ownership and management model, public or private. Alternatively, a municipality with only a few thousand people and a widely disperse population, with a low level of coverage and still requiring high investments will almost inevitably, sooner or later, have to merge with the surrounding municipalities or a neighbouring large municipality to gain economies of scale.

The key question is the following: how much does the mayor want or is willing to be directly involved in the process: a lot, little or almost nothing? If he/she is prepared to participate actively in the process and have conditions to find, internally or in the market, a project leader, with technical and management skills, the creation of a municipal or inter-municipal company is probably a good option. If he/she only wishes to control the basic decisions and results, such as tariffs and investment priorities, it is possible to open a tender to select a private company to create a public-private partnership, or grant the system as a concession. In this case, it is essential that he/she is well assessed to ensure that the tendering process and contract will adequately protect the current and future interests of the municipality and water utility customers. In this context, even taking into account some general principles of the kind outlined above, the decision of the Mayor about what to do with his/her municipal water services will always be difficult as, indeed, the big decisions always are.

### 4.2.4 Economies of scale and political options

The water sector has very clear economies of scale: the capital and operating costs of the water services decrease significantly as the dimensions of the systems increase. These economies of scale are more intense in the smaller water utilities, where even comparatively small increases in the population served lead to significant reductions of the unit capital and operation costs (Figure 4.6). These economies of scale must be

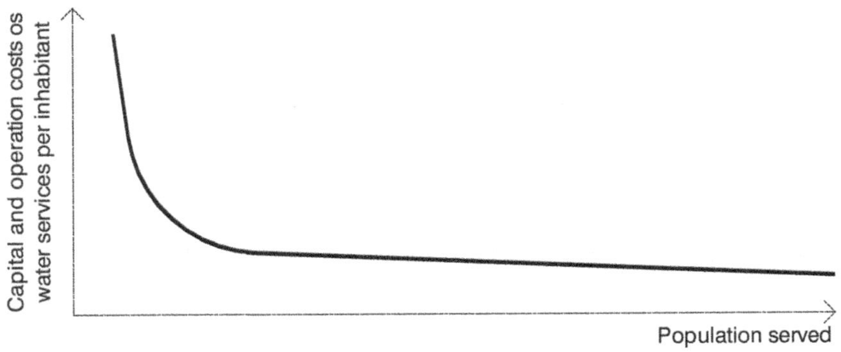


**Figure 4.6** Economies of scale of water services.

taken into consideration in the restructuring of very small water utilities, as it may be virtually impossible to define solutions for them that are technically sound and economically sustainable.

Alternatively, also on sustainability grounds, it is important to be realistic when dealing with situations of a high dispersion of housing: if the individual systems operate satisfactorily, a good solution might be to keep them and, if necessary, organize regional services to take care of their operation and maintenance.

Size and sustainability, in addition to the institutional models, of course, are essential aspects to consider in the early stages of the restructuring processes of water utilities:

- Larger systems may usually adopt any of the legally available models, public or private;
- Smaller systems may have to merge with others to be sustainable;
- Bulk water supply (abstraction and treatment) may be shared by a number of distribution systems;
- There are many advantages for water supply, wastewater and storm water systems to be managed together (integrated management of the urban water cycle).

Basically, the most important decisions regarding the restructuring of water utilities are political and must be made at an early stage. Experience shows that the costs of persisting with a change management process without prior political definition of a sustainability boundary for the water utility may be just too high.

### 4.2.5 The water systems of the future

In the future, raw water will be scarcer, people will value water more highly; energy will certainly be much more expensive and will have a higher influence on water utilities budgets. Therefore, producing and distributing potable water will be more expensive, and tariffs will be higher, even after the necessary improvements on the inefficiency of some current systems.

Alternatively, technological developments will probably lead to lower energy consumption per cubic meter of desalinated water or reused treated wastewater. There has been a significant reduction of the economically feasible minimum dimension of membrane filtration treatment units, which will probably reach the building and individual family size.

With all the limitations of any forecast of this nature, when considering change it is important to think of the water systems of the future, say in fifty or one hundred years. Indeed, this is not too long, considering the useful life of the assets involved, where there are examples of water pipes in operation for more than two centuries.

Some of the characteristics of future public water systems will probably be as follows:

- The systems will not leak (Dellow D. (2012). Personal Communication), they will have less bursts and their operating pressure can be higher, because leakage will be much less pressure dependent;
- The pipes will last well over one hundred years, therefore the current replace/repair dilemma will no longer exist: old pipes should be replaced as soon as possible, purely on economic grounds;
- In new urban developments, pipes will be more commonly installed in galleries permitting visible inspection and easier maintenance operations.

However, perhaps even more radical alterations may take place:

- Much more recycling treatment and reuse within buildings will reduce the use of raw water dramatically along with the volume of sewage to be disposed. Therefore, existing public systems will not need to be replaced due to lack of capacity, and new ones will be smaller;
- In some cases, the use of rainwater collected and stored in buildings, in addition to local recycling of wastewater, may significantly reduce the role of the public systems;
- Local recycling means saving water but, at least equally important, saving energy: the heat from the water we currently use in showers and washing machines will be recovered and not wasted, as almost always happens now;
- There will be fewer financial resources required for construction and operation of public systems (in the streets) and more for private ones (in buildings);
- Jobs in the water industry will be different and will require different skills;
- Technological development and innovation in many areas related to water utility construction and operation, namely about data processing and communication fields, will certainly have a positive impact on the resolution of future problems as it has always happened in the past, helping to overcome catastrophist scenarios.

However, for many years, as changes take place, the existing systems will continue to be partially operated in the same way as they are today.

### 4.2.6 Integrated management of the urban water cycle

The urban water cycle includes water supply, sewerage and stormwater systems, urban streams, rivers and lakes and, in coastal areas, also the beaches (Figure 4.7). Integrated management, either in the same organization or under closely coordinated management of different organizations, has many advantages from the technical, economic and environmental points of view.

There are interfaces between man-made and natural systems that are usually managed by different entities. Water supply and wastewater are often split between different utilities and the same often happens with bulk water supply and distribution and sewerage and stormwater, for instance. All of these interfaces have to be managed: even within the same organization, interface management is always complex and this complexity is certainly higher if there are different organizations involved. In particular, there are clear advantages in integrating wastewater and stormwater systems: the construction and operation of both is similar, the pipes usually share the same or adjacent trenches and because there is always the possibility of illegal discharges and cross connections between systems, the integrated operation of both systems allows for

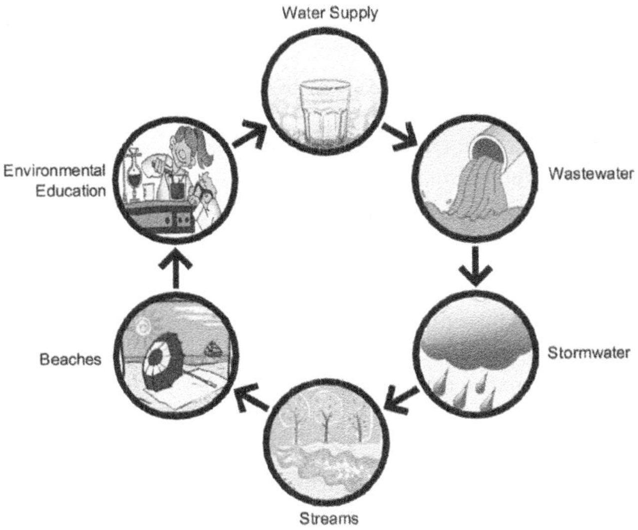


**Figure 4.7** The urban water cycle.

the dry weather collection and treatment of polluted stormwater flows. Placing water quality management of urban streams and beaches under the same organization closes the urban water cycle and allows for their faster and more cost-effective rehabilitation.

However, there are components of the water supply and wastewater systems which have very large economies of scale, namely abstraction and potable water treatment and wastewater sludge treatment. Therefore, creation of regional bulk water supply companies serving a number of municipal water supply distribution systems may have undeniable benefits in terms of reliability and quality of potable water. Also, creation of regional sludge treatment and disposal companies that receive sludge from small and medium wastewater treatment plants has been shown to lead to high economies of scale. The more extensive reuse of treated wastewaters will contribute to reducing the volumes of raw water abstracted and treated water discharged; this is certainly beneficial for rivers that flow in densely populated areas (Figure 4.8).

## 4.3 SPECIAL RESTRUCTURING RECOMMENDATIONS

### 4.3.1 Modernizing old organizations

Typically, water companies have adopted a highly hierarchical structure, with many decision-making levels. They have very complex organizational charts and detailed functions and job descriptions. They operate as a pyramid, with the management board at the apex, several intermediate levels of decision making, and a base of rank and file workers in contact with clients and infrastructure.

There are excessively rigid separations between technical, administrative and commercial departments, which hinder resolution of problems that require interdisciplinary collaboration, such as NRW reduction. The prevailing communication is frequently very formal, sometimes still on paper that circulates around the organization taking days or even weeks on each desk.

This complex and hierarchical organization is certainly not compatible with a quick and timely response to most of the problems that occur every day in water utilities. Also, it does not use the valuable information collected by the employees who have direct contact with clients and the infrastructure. These employees

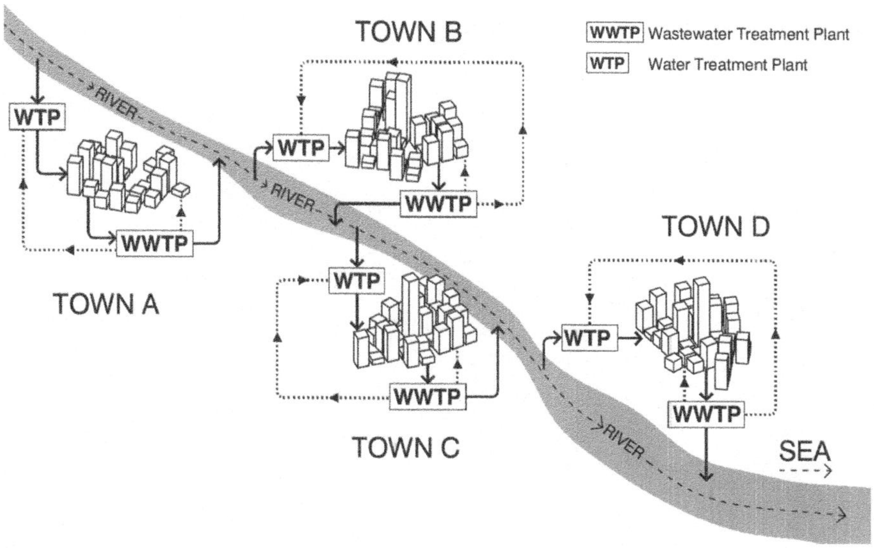


**Figure 4.8** Recycling and reuse perspectives for the urban water cycle.

are not motivated to report the problems that they detect at an early stage, and neither are they empowered to solve them at this stage, when they would otherwise be quicker, easier and less expensive to solve and certainly less damaging to the image and finances of the utility.

Modernizing an old company means replacing everything that is obsolete, from the organizational chart to the computers, operational vehicles, water meters and working clothes. However, it also means changing attitudes and the behaviour of people at all levels in the organization, which is much more difficult to achieve. The selection and motivation of directors and middle managers is obviously of extreme importance. However, it is at least as important, but less frequently perceived, to invest in the front line people who contact directly with customers, such as meter readers, front office employees, telephone operators and with the infrastructure, namely network technicians and active leakage control operators. It is important to recognize the importance of these functions through training, rewarding, trusting and empowering.

The empowerment of the base level of a traditional hierarchical pyramid is very demanding from the organizational point of view and very stressing to middle managers, but it certainly can make a difference in terms of water company performance and how it is seen by their most important stakeholders: the customers.

The water utilities of the past were highly hierarchical, with many levels of decision, and had complex organizational charts and relatively simple jobs. Modern companies, however, have much simpler organizational charts, with few levels of decision but inevitably more complex, demanding and creative jobs. A typical example of coordination, in the past, would be a rowing team (Figure 4.9), with all the members acting together under the command of a voice to move the boat ahead at maximum speed. If they achieved perfect coordination of movement, the power that they transmitted to the boat would be the sum of the maximum power that each member of the team could produce.

However, there are more effective ways to organize teams in which the work actually developed by the group is much higher than the sum of individual efforts (Carlzon, 1989). This means decentralizing and delegating, giving a greater degree of autonomy to employees, namely to the frontline collaborators in contact with customers. This form of organization is more similar to a colony of ants, that derives much of its strength from the fact that each member acts individually, with high efficiency under very sophisticated and effective collective coordination and intelligence (Figure 4.10).

**Figure 4.9** Example of central coordination.

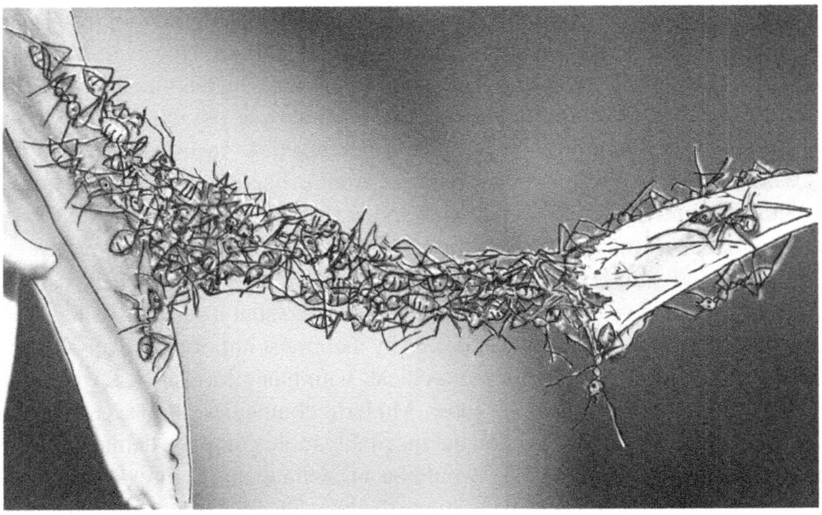

**Figure 4.10** Example of collective intelligence.

It is surprising how much knowledge, experience and energy rests unused in an organization because the leaders no not stimulate their mobilization to the benefit of the organization. It is possible to find workers in water utilities who have been there for many years and have profound experience in how to run the network and even in how the company should operate in other aspects. It is a mistake using only their arms and not taking advantage of their brains, which they are more than willing to also put at the service of the company.

**BOX 4.1 A CENTRAL CONTROL ROOM MAKES THE DIFFERENCE**

Water companies need to pay high attention to their information and communication systems, and related procedures.

Information that is received in real-time should be concentrated in one place, where there should be permanent collaborators trained and empowered to communicate directly with the teams in the field and trigger corrective routine operations. The aim is to resolve the most urgent problems, such as bursts or interruptions of supply, almost immediately.

The Central Control Room provides a more effective coordination of services, higher speed of response, not only in emergencies but also to deal with daily customer requests and complaints.

### 4.3.2 Making water utilities profitable

Private companies are intended to be profitable as a matter of survival. However, some public companies, including water utilities, may systematically close their annual accounts with negative results or receive subsidies to avoid this situation.

In many cases, utilities have charged tariffs that do not meet the full cost recovery principle; hence, costs are partially supported by taxpayers, and not by customers. Where the essential costs are not covered, in addition to the lack of sustainability of the water utility there will be implications in terms of service quality with impact for customers.

To make a monopolistic water company profitable, it would be sufficient to increase tariffs. However, increasing tariffs to cover inefficiency is not ethically correct and may generate adverse public reactions. Therefore, management should review every aspect of the cost and revenue components of the budget first, leaving tariffs to the end.

**BOX 4.2 TRANSPARENCY TO CUSTOMERS AND THE MEDIA: SEE IT ON THE NET!**

It is very easy to include a simple real-time listing of service anomalies and customer complaints chart on the water utility's website, such as the type used at *Águas do Porto*. Abnormal important situations and customer complaints are registered as they occur and, therefore, it may be assumed that everything else should be running normally in the water utility.

This transparency of information is important for customers and the media, creating increased pressure on workers to fix anomalies fast and see them removed from the website listing.

The website helps to implement the rule *"do not leave for tomorrow what you can do today"*, which eliminates most of the reasons for customers dissatisfaction.

It is important to work both on the cost side, searching for efficiency gains in every area of the company and on the revenue side, billing and collecting for all services provided. The potential for efficiency gains in water utilities is mainly possible through the reduction of NRW. Other cost-cutting measures are efficient energy use and reduction of labour costs. The quickest way of saving energy costs is simply avoiding consumption during peak periods, when tariffs are much higher. From personal experience in a bulk water supply company, this measure alone allowed for savings of 10% in energy costs, for the same level of water consumption.

Another measure to save energy in water supply companies is auditing pumping stations and repairing or replacing the pumps that show a reduction of efficiency. Considering that annual energy costs associated with water pumping may be of the same order of magnitude as the cost of the pumps themselves and that the cost of repairing them is much lower, if the measured efficiency of a single pump decreases even by a small amount, say by 5%, it makes economic sense to repair it immediately.

There are, of course, other energy costs that may be reduced through a formal energy consumption audit. This audit is likely to be very cost effective and, as it requires specialized knowledge and equipment, it should normally be contracted externally.

Labour is typically one of the major costs of any water company. In a new water company, it is likely that many activities and services are outsourced and that the number of employees is at the low end of the range of international performance indicators, between 1 and 2 employees/1000 customers. In existing public water utilities, this figure will most certainly be higher: a sign of overstaffing.

On the revenue side, it is important to reduce apparent losses, principally customer under-reading meter errors and un-authorized consumption, increasing the number of paying customers, and billing for all services provided and recovering customer debts.

In many systems, it will be possible to extend the network to serve more clients, outwards to the city boundaries, but these extensions have to be carefully evaluated for cost-effectiveness.

In many cases, companies do not bill all services provided, for example to reconnect a meter that was disconnected due to non-payment or to unblock internal drains of building networks. Beyond the deterrent effect, billing of these services is an additional source of revenue, avoiding costs that are unfairly incorporated into the bills of all customers.

However, is not sufficient just to bill customers, it is also essential to collect the respective payment. Non-paid bills essentially mean a loss of revenue and they are indirectly charged to the paying customers, through tariffs. The collection of unpaid bills, service interruption to non-paying customers (when legally permitted) and the detection and punishment of unauthorized consumption are essential to maximizing revenue and, therefore, to reduce tariffs.

These revenue anomalies should preferably be concentrated in a special unit within the water utility for effectiveness and impartiality reasons: problematic customers are usually a small percentage of the total, most situations are repetitive and it is important, within the law, to treat the same problems in the same way. A specialized unit like this, as will be seen later, in the *Águas do Porto* case study, may also contribute to the early detection of situations of poverty requiring the intervention of the social security services.

### 4.3.3 Dealing with overstaffing

Usually, an inefficient water company has an excessive number of employees. The introduction of efficiency measures worsens the overstaffing problem and the first obvious solution would be the dismissal of the unnecessary workers, but in addition to the social problems this would generate, in public water utilities labour mobility is difficult or even impossible by law. Given the impossibility of reducing the number of employees, part of the solution is eliminating external outsourcing contracts, even if these make economic sense. The corresponding tasks should then be undertaken internally, which requires training of existing workers.

Another alternative is attracting more activities related to the urban water cycle to the water utility, namely management of stormwater systems and rehabilitation of urban streams and beaches.

It is also possible to use existing water utility capacity to solve difficult problems 24 hours a day, 365 days a year, in order to undertake other functions within the city. The engineering department, for instance, may also be responsible for managing or directing other municipal works or contracts, and the central control room may also handle emergencies not related to water.

### 4.3.4 Design and supervision of new works

Many water companies direct a significant part of their financial capacity towards construction of new infrastructure. Planning and design activities are usually undertaken externally by consultants and construction work by contractors. Systems were usually designed to be operated with very little flexibility, essentially in an on-off mode.

Modern systems are different. They must be designed to be flexible, efficient and robust, which requires a different approach and a higher participation of water utility collaborators at the planning, design, construction and operation phases. The network must be seen as a way to deliver water services, not as an end in itself. Each pipe has a normal level or function when in operation but this may be altered in case of emergencies: for instance in the case of bursts, when it should be possible to redirect flows elsewhere to minimize the impact on customers.

To manage a large number of water supply and/or wastewater construction contracts, it is important to create a project management office (PMO) in the water utility, tasked with managing and supervising design and construction works. It should keep the schedules and budgets under control, whilst also ensuring that the knowledge of the operational staff is incorporated in the new works. A well aligned PMO should be capable of defining metrics that can be directly linked to the strategic objectives outlined by management, in order to develop tools and techniques that can give a complete picture of the overall status and compare performance indexes over time. This can bring new levels of transparency to executive management (Project Management Institute (PMI), 2014). In the early stages of restructuring of poorly managed water utilities, this may be difficult due to the lack of reliable, organized historical data. Furthermore, project management methodologies and capabilities are not usually present in the existing organization; therefore, it may be necessary to bring in new people for the job or to contract out the new project management function.

It is necessary to perform the design, procurement and construction phases simultaneously for many different types of projects with many different scopes and locations. Information produced during the design phase should be communicated to team members involved in procurement and construction phases and validated by staff in charge of network operation. In particular, it is not possible to design a modern water supply or wastewater system without local operation experience and without access to the corporate GIS and customer and infrastructure databases.

An aligned PMO should be capable of eliminating long reports, bureaucracy, unnecessary meetings and to communicate to the management team the right information needed to show the entire picture that permits them to remove roadblocks and interfere in the cases that need direct intervention. It is important to be able to display the ongoing projects, in their different phases, at any future date to anticipate potential traffic conflicts and to facilitate the coordination with other public works going on in the city. The integrated management of construction contracts can have a significant impact on the duration of the works and the respective costs. In the *Águas do Porto* case study, for instance, creation of a PMO resulted in a very substantial reduction in both construction times and costs.

### 4.3.5 Increasing reliability of water supply systems

Reliability issues are a major concern for the design and operation of modern water supply systems. The reliability of a public water system can be measured by the likelihood that its customers will be supplied under acceptable conditions. These conditions can be defined for instance in terms of water quality, limit pressures (maximum and minimum) and interruptions of supply (Pocas-Martins et al., 2007).

In the absence of universally accepted indexes, water supply systems reliability is usually measured using simple indexes, which evaluate excess capacity in the network's key components. This approach does not distinguish between different possible failure modes and does not reveal the existence of different performance levels across the network. Routine and catastrophic situations must be dealt with separately (Figure 4.11). The former may be addressed using probabilistic analysis and the latter through the duplication of key components in the network. Pipe bursts and power shortages are examples of routine events. These events are quite predictable because it is easy for water companies to gather data regarding previous events in their operational records. Extreme events include earthquakes, severe floods or droughts, pollution and war, amongst others. Since the occurrence probability of these kind of events is very low, useful data based on previous events is usually unavailable.

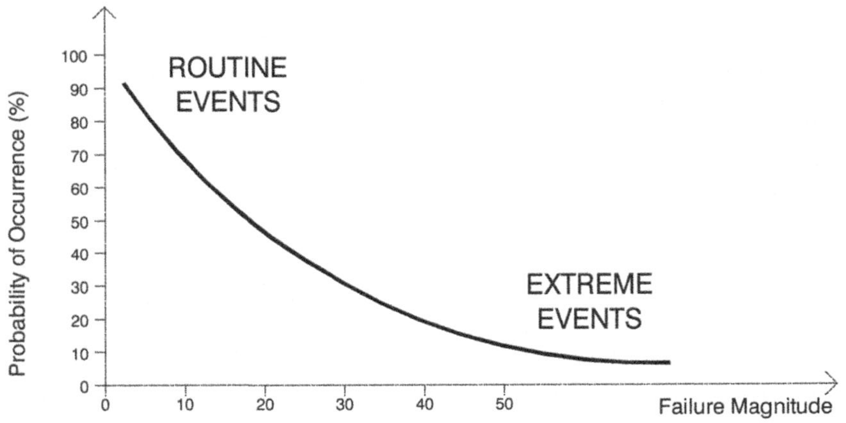


**Figure 4.11** Extreme and routine events in water utilities.

The magnitudes of failures associated with each of these events are obviously different. As a result of these events, supply failures of different durations are expected to occur. Besides the (real) impact on service quality, the possible impact of the event itself and its influence on customer perception of the supply failure should also be considered. As noted, causes of supply failure may range from a simple pipe burst to a natural catastrophe.

Since routine and extreme events have distinct causes, their effects, both real and perceived, are different and the amount of data available for estimation of their probability of occurrence are so dissimilar that each of these kinds of events should be analysed separately.

From a management perspective, the network's vulnerability to different kinds of events should be analysed in order to decide how much should be spent and how in order to increase its overall reliability. Costs related to extreme events, essentially fixed costs, include those necessary to build redundant components for the major parts of the system. Network reliability regarding routine events can be increased almost indefinitely as a result of increasing investment in leakage detection and pipe burst reparation, whilst also through adding redundancy components, in particular by building loops in the network.

Management must, therefore, decide how much money should be concentrated on the prevention of short, recurring service interruptions, to which the customers are particularly sensitive and how much should be spent to mitigate the consequences of a hypothetical extreme event that the customers are unlikely to ever experience.

Information about past behaviour is important for reliability assessment although it should be used with caution. The most popular reliability index for urban water networks, IWA's indicator QS13 (Alegre et al., 2000; Duarte et al., 2004) is based on this kind of information:

$$QS13 = \frac{NI \times DI}{NT \times HY}$$

- NI – Number of customers affected by service interruptions during the reporting period.
- DI – Duration of service interruption, measured in hours.
- NT – Total number of customers.
- NY – Number of hours in the reporting period, typically a year.

Electrical networks also use the information about past behaviour as a measure of reliability. The IEEE 1366 standard includes several indexes of this nature (Warren, 2002). The $CEMI_n$ index (*Customers Experiencing Multiple Interruptions*), for instance, displays some network vulnerabilities that are hidden by indexes that measure a system as a whole:

$$CEMI_n = \frac{CN_{(k>n)}}{N_T}$$

- $CN_{(k>n)}$ – Total number of customers that experienced more than $n$ sustained service interruptions (with a specified duration, for example, one or two hours for water utilities).
- $N_T$ – Total number of customers.

The application of a similar index to water supply networks is believed to be beneficial since the distribution of failures across a network tends to be very heterogeneous.

The disadvantage of this kind of index is that, in general, reliable and auditable information about the duration of service interruptions and the exact number of customers affected by them are not collected by water companies. The actual time of the failure occurrence and respective service reestablishment should be recorded. The consistency of this information can be verified against the operational data gathered by the companies, especially if ERP systems (*Enterprise Resource Planning*) are used to support human resource management.

The number of customers affected can be obtained by two different methods:

1. The simplest method, but also the least reliable and least auditable one, is performed *in situ* by the work team responsible for supply reestablishment, in the cases where the interruption is caused by a pipe burst. This information should be added to the worksheet that is used in many companies to record the time necessary to perform the repair.
2. Given the failure location, the valves that should be closed during the repair can be identified and the number of customers affected can be determined. This method obviously requires a reliable GIS and customer database, but it can be automated, making it auditable. This process is likely to be the most suitable for most water companies, although it requires the development of adequate information systems.

Larger water utilities typically have more than one abstraction/water treatment plant. This makes it possible to interconnect them and always have an alternative if there is a major pollution problem or a failure at a treatment plant. In smaller systems, however, there may be only one source of water and only one water treatment plant. Such systems are vulnerable to major failures of the production of potable water especially if these failures are severe enough to last for more than the usual one or two days storage capacity of the typical service water reservoirs.

It is clear that every water distribution system, large or small, should have at least two alternative sources of supply, independent from each other. In systems with only one source, the least cost solution to providing the required redundancy may be connecting the sources of neighbouring systems so that each of them acts as a back-up to the other. In the Greater Porto area in northern Portugal, shown in Figure 4.12,

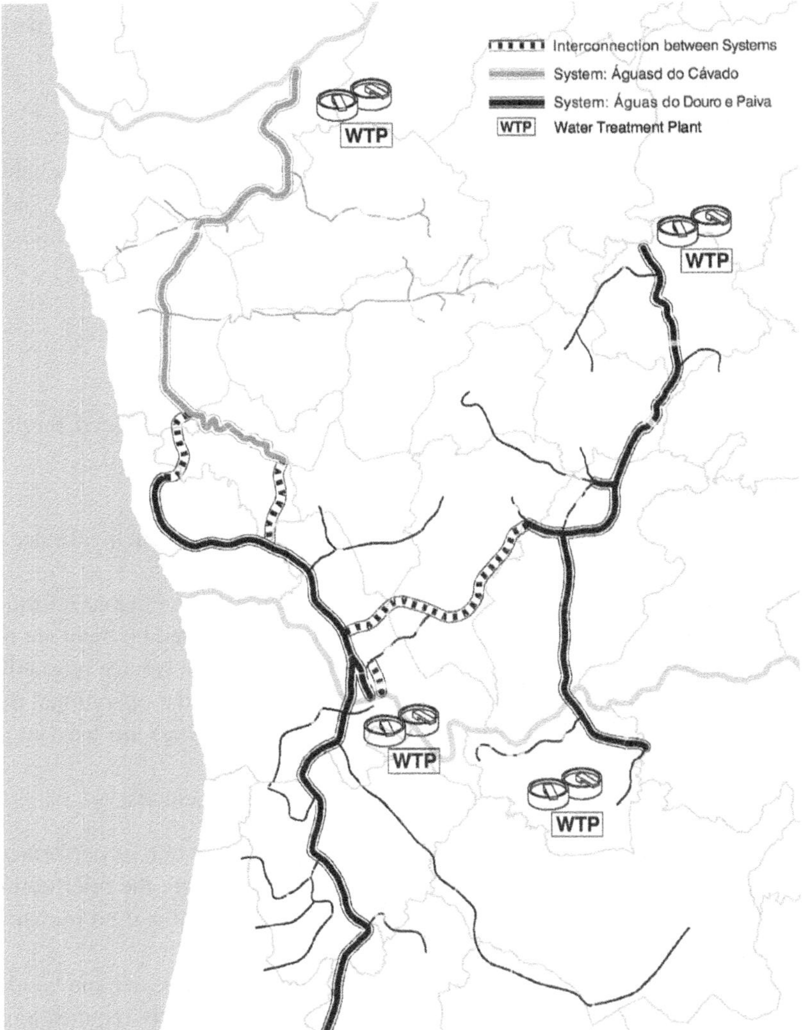


**Figure 4.12** Example of "water highways" between neighbouring water supply systems.

**BOX 4.3 THE IMPORTANCE OF GIS**

It is impossible nowadays to manage a modern water utility without a corporate GIS.

The most obvious use of GIS in a water utility is to model and display the infrastructure database. However, it is also important to collect the coordinates (x, y, z) of each customer (x and y through the postal address and z through the floor number) and geographically link both databases. The much less used z coordinate, still seldom stored in utilities' databases, measures the elevation of the apartment above the ground level and it is important to estimate the available pressure under different conditions, namely during the night (maximum pressure) and at peak flow (minimum pressure).

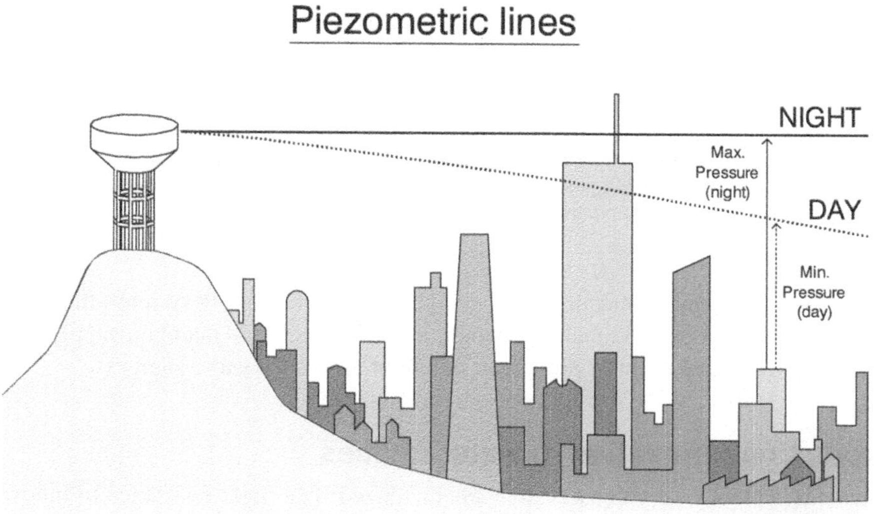


In addition to producing simple maps showing bursts, diameters, materials and age of the pipes or the clients with debt, linking the infrastructure and customer databases allows, for example, to identify those customers with insufficient pressure, who were affected by more than one interruption of service in the last six months or who may be affected by the shutdown of a water treatment plant.

GIS is a powerful tool to help reduce NRW. The measures that are taken to reduce NRW are typically obtained through GIS based algorithms and are presented as GIS produced maps, for instance showing the priority pipes and water meters to be replaced, the zones with excessive or insufficient pressure and the most likely customers with unauthorized consumption.

**BOX 4.4 THE REPLACE OR REPAIR DILEMMA**

Pipe replacement is frequently the first measure that some water companies take to reduce their water losses.

This decision is almost always wrong: not the inevitable decision to replace pipes that have clearly reached the end of their useful service life or have recurrent bursts, but the decision to proceed with this measure first in order to reduce NRW.

Indeed, as the pipes, fittings and connections reach the end of their useful life, burst frequency tends to increase, supply interruptions become more frequent, maintenance costs rise sharply and it may be inevitable to replace them.

(*Continued*)

> However, the age of the pipelines is not, by any means, the only factor influencing the burst rate. In one of the case studies presented in this book, water losses have actually increased after many kilometres of asbestos cement pipes were replaced by polyethylene pipes. The asbestos cement pipes had previously had a history of frequent bursts, but since these were almost always large bursts, they were immediately detected. The new polyethylene pipes, however, had not been installed properly and were poorly welded, which led to higher water losses than the previous bursts in the old asbestos cement pipes which they replaced.
>
> Certainly, for the purpose of NRW reduction, before considering pipe replacement, there are three measures that can be taken to reduce real losses, all without major investment and producing results much more quickly: pressure management, active loss control and quick repair of bursts.
>
> There seems to be a never-ending dilemma in the water utilities world: replace or repair? This has been going on for very long, because when we think we finally know the answers, the questions have changed. In fact, if we were sure we could build leakage free water supply networks that lasted at least a century, the economic decision that made sense would be to replace virtually all the old and suspicious pipes and connections as fast as possible.
>
> However, the last global economic crises made it difficult to increase public debt and borrow money at low interest rates. Therefore, it appears that times are, again in the short term, more inclined towards repairing than replacing. However, without these financial restrictions and if it is possible to depreciate assets over longer periods, closer to their expected life time, the long term optimum solution is replacing the old pipes as fast as possible, even before they start bursting frequently and originating measurable leaks.

the construction of trunk mains connecting three neighbouring water supply systems, the so-called *"water highways project"*, at a cost of €16/inhabitant would make it possible to maintain supply continuity with only minor restrictions, in any event of prolonged failure of any of the water sources.

### 4.3.6 Introducing innovation in water utilities

The water sector is typically very conservative, characterized by a great resistance to innovation. There is an *"I know what I like; I like what I know"* attitude, to some extent understandable considering that ensuring the continuity of the water services is paramount to operational water utility people. However, this over conservative attitude may result in inefficiencies and higher costs.

In recent years there have been important developments in areas such as pipe materials and fittings, new kits to repair bursts without interrupting the supply, variable speed pumps, trenchless pipe installation, automatic meter reading, monitoring, telemetry, mathematical modelling and GIS. New technologies were also made available in the market to detect leaks and bursts, such as ground penetrating radar, continuous noise monitoring of the network, advanced pressure management and expert time series analysis systems that help detect and locate leaks and bursts (Council for Science and Technology, 2009).

Participating in specialized international meetings and fairs and consulting the reference journals is essential for knowing about the most recent technologies and equipment. There is an understandable caution by the majority of water utilities to try the latest technologies, but those approaching excellence are more likely to adopt them at an early stage. To install a new technology in a water utility, it is necessary to have trained, open minded and motivated collaborators and this is often the critical factor. In fact, integrating new technologies and new ways of doing things in an existing operational water utility may lead to higher efficiency or reliability but also involves additional efforts and risks, and certainly it will not happen without the commitment of top management.

Cooperation between water companies and universities through joint applied research and development projects is a great way for water companies to introduce innovation, with benefits for both parties. Sponsoring short-term graduate and Masters projects on a regular basis, with the students working at the water company, under joint supervision of an academic and a company director or water specialist is an excellent way to permanently introduce innovation into water utilities. Alternatively, developers of the latest technologies

**BOX 4.5 TRENCHLESS TECHNOLOGIES**

Trenchless technologies to install, replace or repair water and wastewater pipes have become more and more competitive in terms of construction costs in relation to the conventional methods, which virtually destroy pavements and create traffic chaos in cities.

This is most certainly the solution in urban areas and even in suburban areas for trunk mains and distribution pipes without too many connections.

**BOX 4.6 AMR - AUTOMATIC METER READING**

As prices go down and existing meter readers retire, it will be more and more appealing to install AMR – Automatic Meter Reading-in water utilities.

Data transmission may initially be through simple walk-by or drive-by methods, but it is very likely that in the future, as communication prices go down, customer consumption data is will be received in real time at the Central Control Room of the water utility. This will be facilitated if the water, electricity and gas utilities join efforts to share a common data transmission system, probably using copper wiring or optical fibre networks.

are usually more than willing to test them in reference water utilities. Cooperation between technology developers and water utilities, therefore, is another way of introducing innovation, with mutual advantages.

## 4.3.7 Stakeholder relationship

Today, organizations know that to seek only customer satisfaction is not sufficient in order to achieve success. It is necessary to establish a dynamic relationship that creates value for all stakeholders, encompassing all individuals, groups or institutions that might affect or be affected by the utility's decisions, practices or goals, namely customers, owners, employees, the regulator, suppliers, public authorities and unions (Figure 4.13). It is impossible to meet all of the individual interests of all the different stakeholders. All stakeholders have expectations, but they also should know that their expectations may conflict with those of the others.

Management plays a crucial role in this permanent equilibrium reaching process. It should never be a part of the problem and must be seen by all as being a part of the solution. Transparency and sharing common information certainly help the negotiation process.

### *4.3.7.1 Customers*

Customer satisfaction is a matter of survival to every company in the market. Water utilities are monopolistic, but their customers can present complaints that may be analysed by the regulator, by consumer associations, by the media and by politicians.

It is important to know and thoroughly analyse all complaints and their motives. There are no excuses for neglecting customer satisfaction because, in the majority of cases, doing a good job does not cost more than doing it badly. There must be a zero target for complaints, but no matter how good the service is, there will always be people wishing to complain about the service interruption, lack of pressure or, more often, about a bill or excessive waiting time on the phone or at the customer service desk of the utility.

The regulator usually controls complaints and the manner in which the water utilities deal with complaints, including the time to respond to customers with a substantive written answer or resolve the problem. Facing a complaint, there are only three possibilities: the client is right, partly right or wrong. In

**Figure 4.13** Major stakeholders of water utilities.

any case, a fast, polite answer is an obvious thing to do. In any case, it is important to take every relevant complaint as an opportunity to improve and find solutions to avoid repetitions.

Sometimes, following a growing number of complaints, management may be tempted to decide to allocate more people to customer services. This is usually a wrong measure. In fact, in a well performing water utility, there should be very few complaints, whilst there are no good reasons for a customer coming to the company's contact points.

Accordingly, water utilities must:

- Prevent customer complaints through improvement of service quality, fair tariffs and transparency with timely and effective information about supply interruptions and works, amongst other factors;
- Create conditions to avoid situations where customers have to come to the utility counters to sign or alter contracts or to pay bills, as virtual online contact centres and alternative forms of payment are preferred and can easily be made available;
- Provide a good service when customers decide to phone or to come personally to the counters.

### *4.3.7.2 Owners/shareholders*

The major decisions concerning a water utility are taken by the representatives of the owners, often elected politicians. These representatives have the ultimate responsibility for the water company, even if it has been privatized and they are to be penalised if there are many complaints from customers, most of them who are also voters. They appoint the top management of the water utilities, establish their major objectives and strategic guidelines, including selection of the management model, and may undertake a major role in terms of fixing tariffs and distribution of the annual results.

The positioning of the owners may vary along with the political cycles; management must be able to understand these variations and incorporate them into their own decisions. The top management of the water utility may, of course, influence the owners' representatives on strategic and operational matters, in the best interests of the water company.

Stability is an important value for implementing any strategy in a water utility. A single political cycle is certainly too short for that, therefore a number of solutions have been found to overcome this problem, namely:

- A greater influence for the regulator;
- A private shareholder, typically in a minority position;
- Management contracts with fixed objectives and conditions;
- Privatizing the operation, namely through a concession model.

#### *4.3.7.3 Employees and unions*

Managers sometimes see unions as adversaries and vice versa. Fortunately it is not always the case and, indeed, it is not possible to reach excellence without sufficient motivation and commitment from employees, which requires good working relations. However, management sometimes has to make decisions that have a negative impact on the workers due to legal impositions or instructions from owners or the regulator.

It is only natural that management attempts to prevent workers from asking the unions to solve problems, which could otherwise be avoided or resolved directly with them. In companies with traditional hierarchical structure, the unions mostly represent those employees at the bottom of the pyramid. That is why their function is to examine and question the decisions taken by the company's management on behalf of their members. When the organizational structure becomes flatter, decisions tend to be more transversal, which radically changes the role of the unions.

Another role of unions is in examining how the company complies with labour legislation and collective work agreements whilst meeting with management around the negotiating table on behalf of their members. Creating a permanently open communication channel between the board and workers avoids unnecessary tensions, allows for early awareness of potential problems and creates conditions to discuss and implement corrective measures, avoiding the scaling up to the union level. In any case, it is important to realize that union leaders are professional negotiators. Typically they have a long list of demands

**BOX 4.7 FROM METER READERS TO DISTRIBUTION ZONE MANAGERS**

For the vast majority of customers, the only personal contacts with the water company are established through the meter readers. These meter readers, often working under an outsourcing regime, are given daily itineraries and expected to obtain the essential figures that support the next customer bills, namely the current cumulative consumptions shown in the respective water meters.

To avoid the meter readers "falsifying" the readings without visiting the actual installations, they are often not given the previous readings or, indeed, any information about the customers. They go from meter to meter to collect the figure on which the next customer bills will be based.

From personal experience, a much better alternative is upgrading the job description of the meter readers, transforming them into zone managers. Instead of being given only daily itineraries to follow, they are given a zone to manage, with the responsibility to collect the information that is required for billing but also to locally inspect the customer meters, detecting, for instance, stopped or faulty meters. As zone managers, they are also given access to the historic consumptions of each customer. This helps them to extend special attention to customers with suspicious consumption totals or who have made a recent complaint.

to put on the negotiating table with management, some more important than others. Experience shows that it is almost always possible to resolve them appropriately, in due time.

#### 4.3.7.4 Suppliers

Water utilities contract suppliers in a competitive market; therefore it is expected that the relations with them are smooth and clear. However, problems may occur if the goods and services are not precisely specified in advance.

There are two main reasons for contracting services: when the utility could perform the work internally, but prefers not to do it, or when it does not possess the required capacity or ability. When a job is done internally, many aspects of it do not have to be explicitly defined, but when it is subject to open tenders, it must be specified in detail; otherwise it is most likely that there will be disputes in the future. Common- sense and contract law dictate that clarification, detailing and negotiation should occur in due time, certainly before signing the contract. However, some contracts, mainly longer-term ones may have to be altered for unforeseen reasons, therefore, binding agreements should include specific items to frame and regulate future alterations.

In water utilities, most of the services, equipment and works that go to tender are trivial and repetitive and there is no reason for them not being well specified. Alternatively, in competitive markets, typically in developed countries, there are always many bids to choose from. Experience shows that, under such circumstances, the best option is awarding contracts to the lowest bid, avoiding more complex criteria. Indeed, it is very easy to explain awarding a contract to the lowest bid, but it certainly takes a few pages of explanation and eventual litigation if it is awarded to the second lowest bid, and so on.

The situation is different when a water utility requires a new technologically complex item, for example a water or wastewater treatment plant, in which there are specialized patented technologies involved; in these cases, simple contracts awarded to the lowest bids are certainly not the best option. Nowadays, there are many alternatives for contracting such specialized equipment, including the BOOT – Build, Own, Operate and Transfer-that transfers the technological risks to the contractor.

#### 4.3.7.5 The regulator

The water sector has characteristics that make it a prime target of state regulation. These services are natural monopolies of a local or regional nature, which naturally limits the competition and certainly customers cannot choose between alternative water suppliers. Although the number of countries with water services regulators has been increasing, they are still in a clear minority. Table 4.2 shows the countries with water regulators in 2014, according to the list of invitations to the First World Regulators Forum, scheduled for the 2014 IWA Congress in Lisbon (Baptista 2014, Personal Communication).

The vast majority of water utilities worldwide are publicly owned and it could be argued that a regulator would not be strictly necessary in these cases, as the ultimate responsibility for the quality of these services, even if they are under private management, is political and hence is evaluated in subsequent elections. In fact, there is some controversy when it comes to the regulation of municipal services, as it may be regarded as an interference of a non-elected body in the legal attributions of the municipalities.

In principle, the regulator should protect not only the interests of customers in terms of quality of service and tariffs, but also the sustainability of the water utilities. Regulatory agencies can be entrusted with various responsibilities, including tariff approval, assessment of service quality through performance indicators and benchmarking and supervision of private operator contracts. Sometimes, they also have the power to deal with customer complaints, arbitrate contractual disputes, monitor performance of water companies and evaluate documents that lead to the granting of water services. Regulators are considered to be independent if they

**Table 4.2** Countries with water regulators in 2014 (Source: Baptista 2014, Personal Communication).

| Europe | Africa | Asia | America | Oceania |
|---|---|---|---|---|
| Albania | South African Republic | Egypt | USA (40 regulators) | Australia (7 regulators) |
| Armenia | Gambia | United Arab Emirates | Canada (4 regulators) | Papua New Guinea |
| Belgium | Ghana | Philippines | Anguilla | |
| Bulgaria | Cape Verde Islands | Indonesia | Argentina (13 states) | |
| Kazakhstan | Mauritania | Maldives | Bahamas | |
| Scotland | Mozambique | Sri Lanka | Barbados | |
| Slovakia | Niger | Malaysia | Belize | |
| France | Kenya | | Bolivia | |
| England | Ruanda | | Brazil (34 regulators) | |
| Ireland | Tanzania | | Chile | |
| Northern Ireland | Zambia | | Colombia | |
| Italy | Mali | | Costa Rica | |
| Kosovo | Malawi | | Ecuador | |
| Latvia | | | Guiana | |
| Lithuania | | | Honduras | |
| Portugal | | | Jamaica | |
| Romania | | | Nicaragua | |
| Ukraine | | | Panama | |
| Denmark | | | Peru | |
| Malta | | | Trinidad &Tobago | |
| Croatia | | | Uruguay | |
| Hungary | | | Paraguay | |

act in accordance with the law, keeping equidistant in relation to private and political interests. However, this model is not necessarily effective if political intervention compromises regulatory independence.

The relation between the regulator and management of water utilities is defined by law and tends to be very formal. It might be said that, when regulators are doing their job properly, they should be "hated" by managers of water utilities. In fact, the regulator is expected to evaluate and put pressure on water utility management in order to improve their performance and customers relations. Managers obviously know their water utilities better, but the regulator has information on many water utilities and both have to justify their positions and assumptions, and act within the same legal framework. Overall, good professional managers have nothing to hide or fear from good regulators. It is in the best interest of both that the water utility has high quality of service and satisfied customers, but also that the water utility is sustainable.

There are potentially more conflicts between the owners and the regulator, as they may give contradictory instructions to the management of the water utilities. The elected representatives of municipalities would certainly want to have the final word regarding strategic decisions concerning their water utilities, namely annual approval of tariffs and definition of the tariff structure, such as the eventual differentiation between types of users and between smaller and larger consumptions, which has a great social impact. Alternatively, if the regulator has been given that responsibility by law, it will tend to set a similar tariff structure for all the regulated water utilities and to approve the annual tariffs according to its own criteria.

#### 4.3.7.6 The media

In respect of the media, the proverb *"no news, good news"* certainly applies for mature water companies.

In developed countries, water services tend not to be an issue because they have been taken for granted for many years. The exceptions are usually bad news: big bursts, exceptional service interruptions or high tariff increases. In water companies that are undergoing a change management process, however, there are more opportunities to have positive news in the media: system expansion may be occurring at a faster speed than anticipated, water loss might have decreased significantly or ingenious water thefts have been discovered.

Understandably, water utility management would like to publicize their strategic projects and successful achievements, but it is ultimately up to journalists to decide what to publish or broadcast. From personal experience, it is possible to reach win-win situations with the media through provision of material that leads to good news for both the public and the water utility.

In fact, occasionally in every water utility there are events or results which are sufficiently interesting so as to be made known to the public. Bringing these specific issues to journalists' attention is also an opportunity to inform the public on the more general plans and strategic projects of the water utility, as news usually goes from particular to general issues, not the other way.

The major problem of water services in terms of the media is that most water works are underground and cannot be seen and, when they are visible during the construction phase, they are mostly associated with a negative impact on city life.

It was discussed above that there are many technical and economic advantages of expanding water utility activity to include the whole urban water cycle, namely rehabilitation of urban streams and beaches. In fact, there is another, less obvious, advantage: it is through these activities, which are more likely to attract public and media attention, that many people realize the relevance of their water utility.

#### 4.3.7.7 ENGOs

Environmental non-governmental organizations (ENGOs) are essential in society as they are concerned for the environment in the long run, without having to make short-term compromises. They are represented on a number of advisory bodies and are often asked by the media to comment and eventually contradict official positions. Some tend to support fundamentalist positions; others share the table with authorities, including water utilities and participate in the discussion of issues related to the potential long-term environmental and development debate.

Water utilities and environmentalists are natural allies in many respects: for example, both want full territorial coverage with modern and well operated wastewater systems so that streams, rivers and the sea are as pollution free as possible. In particular, there is an ENGO that has an historical role in the protection and rehabilitation of the coastal areas of Europe and the world, the Blue Flag organization, which has, for the last twenty years been classifying the beaches and rewarding the best with a symbol of excellence that is internationally recognized. People select Blue Flag beaches for their holidays and politicians are pressed to obtain such valuable quality status.

## 4.4 A SIMPLIFIED RESTRUCTURING MODEL

Mature and economically sustainable water utilities tend to present profits through tariffs that ensure full cost recovery, although in the past many of them may have used transfers, both national and international subsidies, to partially finance the investment in infrastructure. However, the vast majority of global water utilities, unfortunately, have not reached this sustainability stage and have still to invest a significant amount of money to serve more customers and to improve coverage, quality of service, reliability and efficiency.

The task is easier, of course, for a large utility that is expanding its intervention area and more difficult in a small system that has few paying customers. There are cases in which the amount of NRW is very high and its reduction may be sufficient to avoid tariff increases. Alternatively, in very small systems requiring high investment, a local sustainable and affordable solution may be practically impossible.

Water utilities are supposed to endure for many years, because the service they provide is essential, and the useful life of their assets is of the order of magnitude of many decades. The corresponding economic studies and financing solutions must be also long term, but experience shows that it is important to avoid over optimistic assumptions about the future: too many customers, too high per capita consumptions, too high tariffs and an unlimited borrowing capacity at low interest rates.

In some cases, there is a window of opportunity to support the expansion of the infrastructure, benefiting from lost fund subsidies, but decision makers tend to forget that there is seldom any kind of support for operating costs and the availability of such subsidies may result in constructing excessively large systems, for instance to accommodate existing high values of water losses in the distribution systems.

The well-established TTT (Tariffs, Transfers and Taxes) economic sustainability model applies to any water utility. However, its practical application in existing water utilities may be supported by the simplified restructuring model presented in Figure 4.14, which has been successfully tested in real water utilities.

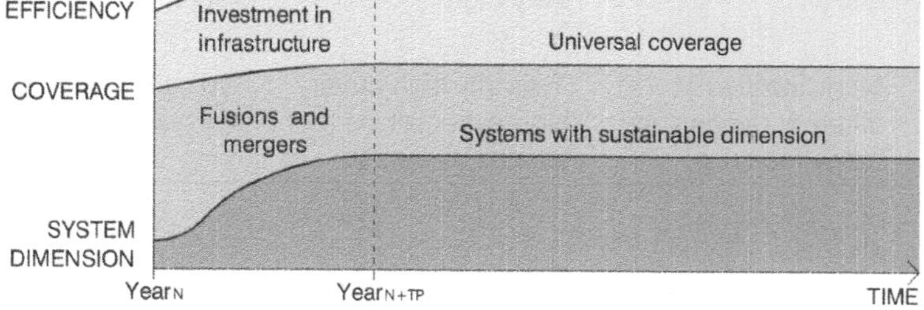


**Figure 4.14** A simplified water utilities restructuring model.

Departing from the existing situation, it is essential to assume that restructuring is going to take place over a transition period characterized by a relatively high organizational turbulence. During this period, it is usually possible and desirable to progress simultaneously in four different directions from the beginning: (1) redefining the sustainability boundaries of the water utility; (2) increasing efficiency; (3) investing in the expansion and rehabilitation of the systems; and (4) adjusting tariffs.

These lines of action should be followed with the appropriate degrees of intensity and converge on sustainable objectives at the end of the transition period, as follows:

(1) If the system is too small, it is essential to redefine its sustainability boundaries. There are cases of sustainable very small systems in isolated places, benefiting from virtually pristine water sources, but these are, unfortunately, the exception to the rule. If there are too many small unsustainable systems at regional or national level, it is necessary to reduce the number of systems through fusions or mergers. Investing in system restructuring without future sustainability perspectives is, unfortunately, a very costly and common mistake.

(2) In any water utility there are always opportunities to increase efficiency, including reducing NRW, energy consumption and labour, which should always be priority issues.

(3) In most cases it may be necessary to invest in system expansion in order to serve more customers or rehabilitate existing infrastructure. These investments, however, must take into consideration the potential NRW reduction, the possible simplification of the networks, by for instance reducing the areas served by pumping and the reduction of pressure that may increase the useful life of the infrastructure and also contribute to NRW reduction.

(4) Finally, with everything else defined, it is possible to calculate the full cost recovery tariff and to establish a sensible transition towards this goal.

It is interesting to analyse the change management processes of *Águas do Porto* and *Águas de Gaia* and a possible sustainable solution for a hypothetical country or region with many small municipalities using this simplified restructuring model (Table 4.3).

**Table 4.3** Application of the simplified restructuring model (initial conditions).

| | Dimension | NRW | Investment required | Tariffs |
|---|---|---|---|---|
| *Águas do Porto* (1997 – 2004) | OK | Very high | Relatively low | OK |
| *Águas de Gaia* (2006 – 2013) | OK | Very high | Very high | OK |
| Country or region with many small municipalities | Municipalities of different dimensions, many too small | Typically high or very high especially in the smaller systems | Typically high or very high | Typically too low in the smaller systems |

The case study of *Águas do Porto* (Figure 4.15) is typical of a mature but previously very inefficient water utility, with half a million equivalent population. It was possible in a very short period of time to achieve great savings through the reduction of NRW. These savings allowed expansion of water utility activity areas to cover the whole urban water cycle, to complete the wastewater system and rehabilitate part of the old water supply system without increasing the tariffs in real terms (there were tariff rises, but below inflation). It was also possible to achieve important efficiency gains in terms of energy and manpower: the number of employees was substantially reduced but without dismissals – the retiring employees have not been replaced and the expansion of areas of activity absorbed workers who have been recovered as a result of a more efficient operation.

The case study of *Águas de Gaia* is similar to *Águas do Porto* in terms of ownership (in that they were both owned by their respective municipalities), in terms of number of customers (150,000 in Porto and 120,000 in Gaia) and expansion of the areas of activity to the whole urban water cycle, whilst also in terms of the change management process that has taken place (Figure 4.16). However, in Gaia it was necessary to invest significantly in the expansion of wastewater systems and, to a smaller extent, in the water supply system but, like in Porto, the tariffs also decreased in real terms due to savings made through NRW reduction.

The following case of a country or region with too many small municipal water utilities is typical in many parts of the world (Figure 4.17). It is assumed that many of the systems need a significant investment

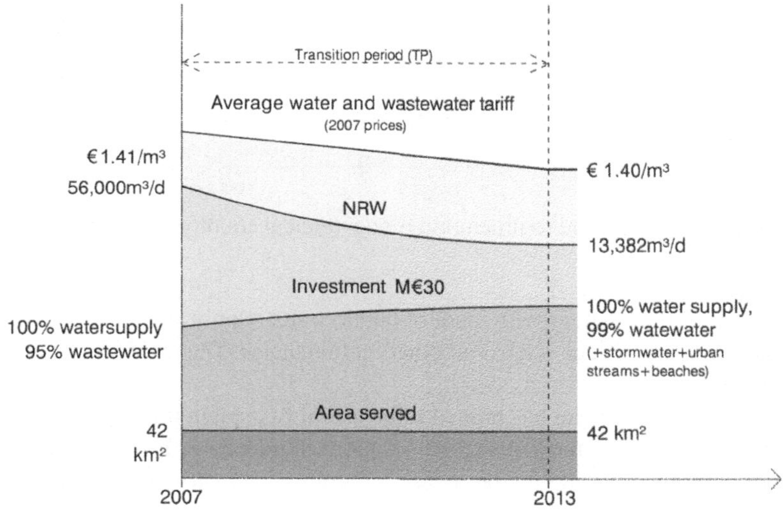


**Figure 4.15** Example of the simplified restructuring model (*Águas do Porto* 2007-2013).

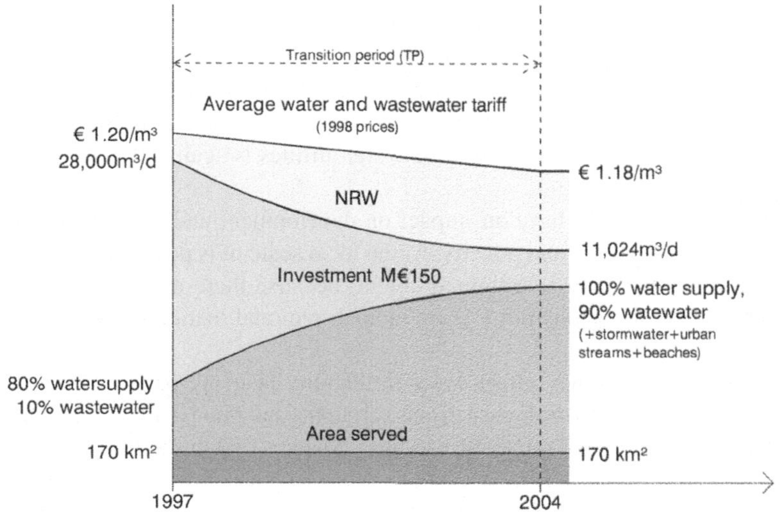


**Figure 4.16** Example of the simplified restructuring model (*Águas de Gaia* 1997-2004).

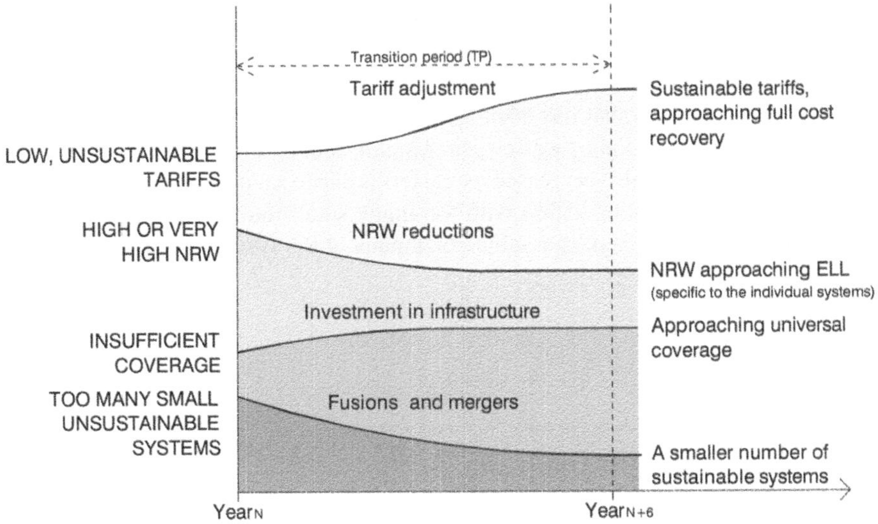


**Figure 4.17** Example of the simplified restructuring model (typical country with many small systems).

to increase the coverage of the territory with modern public water supply and wastewater systems and that there are great opportunities to reduce NRW and other inefficiencies. The tariffs are in most cases too low, far from full cost recovery levels.

This simplified restructuring model has proved to be useful to explain the key aspects to non-technical people involved in change management processes, namely the need for a reasonable transition period, the surprising perspective of NRW reduction as an opportunity to restructure a water utility eventually without tariff increases, the need to consider sustainability dimension of the systems from the beginning of the restructuring processes and evolution towards full cost recovery tariffs under universal coverage, sustainability and efficiency conditions.

## 4.5 CONCLUSION

There are many thousands of water utilities in the world operating under very different conditions. Some are stable and sustainable, mostly in developed countries, many are performing poorly and some are improving. It is possible to compare them using international performance indicators and benchmarking studies but the most important competition of those water utilities is against themselves, for the benefit of their customers.

There are global issues that may have an impact on development and performance of water utilities, namely climate change, demography and poverty. At the local scale, it is possible to mitigate some of these problems in water utilities, but only to a limited extent, because there are practical limits to solidarity. Individual, social and political responsibility at municipal, regional, national and international levels are crucial to find long-term solutions.

Climate change and variability may affect water availability in many parts of the world. The number, magnitude and duration of extreme hydrological events (droughts and floods) affect the quantity and quality of raw water feeding treatment plants and it may be necessary to adapt systems to this new reality: more storage capacity will be needed, alternative water sources and more sophisticated water treatment plants to deal with more frequent, lengthy and severe droughts and floods. This also means higher investments and tariffs.

There is certainly unanimity in agreeing that everyone should have access to water and sanitation for health and reasons of dignity. This is a state obligation but to materialize this objective at a local level, an enhanced level of creativity and commitment by service providers is necessary, along with ENGOs and individuals, to transform the human right to water into a real access to water and sanitation services. Water utilities must participate in the solution, being effective and efficient and eventually applying social tariffs. However, they do not have the means to identify individually the customers with highest needs.

In the 1990s, Public Private Partnerships (PPPs) were seen as a major solution for restructuring poorly performing water utilities, by bringing better management and financial resources for investment in infrastructure. Some experiences have passed the test of time but others have failed. Certainly, the private financing did not match the initial expectations and is more realistic to expect that in the future PPPs are more focused on improving operational efficiency and quality of service, through performance based service contracts and subcontracting.

Some general conclusions may be taken from the water utility restructuring issues addressed in this book. Perhaps the most important one is that, under ideal conditions, a water utility can approach excellence, in four to six years. Reducing NRW may be the key aspect that makes such a change possible. The measures to reduce NRW require the commitment of top management and the participation of virtually everybody in the utility. Water utilities owned entirely by municipal or other public entities are very much dependant on political cycles whilst, sooner or later, they may be subject to decisions that are not in line with the interests of their customers. Stability is a value in itself once a water utility has reached excellence. Therefore, although water is essentially a political issue, in the interest of the customers they serve, it is important to isolate the operation and the technical strategies of the water utilities, as much as possible, from political cycles.

A regulator may contribute to company stability and to customer protection against decisions that may affect service quality or tariffs. Concession contracts, in theory, should also introduce stability, because they usually last for twenty or thirty years, with agreed performance indicators and tariffs. However, they are also vulnerable to the economic environment and political cycles and invariably have to be renegotiated at a later date, which may be very difficult if the original binding documents have not explicitly ruled this situation.

Partial privatization of the capital of a water utility keeps the public control but brings into the Board representatives of the private shareholders that tend to support a professional management and decisions that make economic sense and, consequently, also sustainable customer protection. Should the water utility be listed in the stock exchange, such as the COPASA case study presented in this book, there is an additional external control over its performance, namely through the audited information which is periodically made available. The participation of the private sector in the ownership and management of water utilities is likely to take place under a range of different and more complex models.

There seems to be a trend to reduce the number of concessions with private investment of capital all over the world and, on the other hand, large public water utilities are operating outside their natural limits, acting in fact as private companies. International private investors not previously related to the water utilities sector have acquired well managed public and private water utilities, creating new groups with knowledge of water utilities management and a high capacity of investment. It will be interesting to follow their future role in both the developed and developing countries, eventually associated to local partners to control the political and regulatory risks which have undermined the traditional concessions market in the recent past. It will also interesting to analyse the future developments of the multi-utility public and public-private companies that exist in parts of the world, including some of the following services: water, energy, communications and public transportation.

The separation of the individual functions makes the task of the regulators easier, but those companies present as an advantage the cross-subsidizing possibility, making it easier to guarantee the poorest groups of the population the access to the most essential services at affordable prices.

The following recommendations may be useful for readers who will be involved in change management processes in water utilities.

1. For the Mayor of a medium or large municipality
   - Virtually any institutional model under public or private management is likely to be made successful in medium and large municipal water utilities, therefore the decision is determined mainly from the Mayor's personal preference.
2. For the Mayor of a small municipality
   - Consider merging the system with a larger and well organized neighbouring water utility;
   - Do not risk a concession, especially if the investment required to expand the system is significant;
   - Contract out services to improve efficiency of the system, mainly NRW reduction.
3. For a Mayor who has decided to restructure their water utility under public management
   - Appoint a project leader with previous experience in similar processes and give him/her clear objectives and political support.
4. For a Mayor who has decided to privatize management of their water utility
   - Contact colleagues who have been through this kind of process before;
   - Set the political high level objectives for the concession, including coverage, global investment and maximum tariffs that are socially acceptable and have them tested for economic sustainability;
   - Get the best technical, economic and legal consultants to prepare the bidding process and the contract documents.
5. For someone invited to lead the restructuring process of a water utility
   - Make sure that you trust the representative of the owner and that he/she trusts you;
   - Evaluate the context, the water utility situation and the time available for the change management process;
   - Define realistic objectives for the restructuring process within the existing context and test them with the representative of the owner.
6. For the Project Leader of a restructuring process of a water utility
   - Select the key collaborators very carefully;
   - Contract the best consultants to fill the gaps in your own team's knowledge and experience;
   - Create a promotions and wage differentiation system for the employees based on merit;
   - Introduce a variable component in wages related to objectives and results;
   - Trust your objectives, trust your people and trust yourself.

## 4.6 REFERENCES

Albuquerque, C. (2012). On the Right Track: Good Practices in Realising the Rights to Water and Sanitation. ed. Textype, Lisbon.

Alegre H., Hirner W., Baptista J. M. and Parena R. (2000). Performance Indicators for Water Supply Services. IWA Publishing, London, UK.

Carlzon J. (1989). Moments of Truth. Harper Paperbacks, New York.

Council for Science and Technology (CST) (2009). Improving Innovation in the Water Industry: 21st Century Challenges and Opportunities. Council for Science and Technology, London, UK.

de Mästle C. T. and Izaguirre A. K. (2008). Recent Trends in Private Activity in Infrastructure: What the Shift Away from Risk Means for Policy, PPIAF – Public–Private Infrastructure Advisory Facility, World Bank, Gridlines, Note No. 31, May 2008.

Duarte P., Alegre H. and Baptista J. M. (2004). Performance Indicators for Water Supply Systems (in Portuguese). ERSAR, Lisbon, Portugal.

Gulbenkian Think Tank on Water and the Future of the Humanity (2014). Water and the Future of Humanity: Revisiting Water Security. Calouste Gulbenkian Foundation and Springer, Lisbon, Portugal.

Marin P. (2009). Public–Private Partnerships for Urban Water Utilities: A Review of Experiences in Developing Countries, The World Bank and PPIAF – Public–Private Infrastructure Advisory Facility, Washington, D.C., USA, February 2009.

Marin P. and Izaguirre A. K. (2008). Private Participation in Water: Toward a New Generation of Projects?, PPIAF – Public–Private Infrastructure Advisory Facility, World Bank, Gridlines, Note No. 14, September 2008.

Pinsent Masons (2014). Pinsent Masons Water Yearbook 2012–2013. Pinsent Masons, London, UK.

Pocas-Martins J. and Martins J. P. (2007). Reliability Indicators for Water Supply Systems, Proceedings of GESCOM Conference, Faculty of Engineering, University of Porto, Porto, Portugal.

Project Management Institute (PMI) (2014). The High Cost of Low Performance, Newtown Square, USA.

United Nations (UN) (2008). The Millennium Development Goals, Report 2008, United Nations, New York, NY, USA.

van den Hurk B. and Jacob D. (2009). The art of predicting climate variability and change. In: Climate Change Adaption in the Water Sector, F. Ludvig, P. Kabat, H. van Schaik and M. van der Valk (Ed.), Earthscan, London, UK.

Warren C. A. (2002). Overview of the Full Use Guide on Electric Power Distribution Reliability Indices, Power Engineering Society Summer Meeting, 2002 IEEE, July 25, 2002.

Wild D., Francke C. J., Menzli P. and Schön U. (2007). Water: A Market of the Future. SAM Sustainable Asset Management AG, Zürich, Switzerland.

World Bank (2002). A Sourcebook for Poverty Reduction Strategies. Volume 2: Macroeconomic and Sectoral Approaches, J. Klugman (ed.), Washington, D.C., USA.

World Bank and International Bank for Reconstruction and Development (2008). Global Monitoring Report 2008: MGDs and the Environment – Agenda for Inclusive Sustainable Development. International Bank for Reconstruction and Development and World Bank, Washington, D.C., USA.

# Chapter 5

# Case studies

The following case studies have been selected for their relevance in terms of illustrating key aspects of change management and have been edited to include mostly points of general interest.

The first case study covers a 120 year-old water and wastewater company serving over 150,000 clients, initially very inefficient with around 54% NRW; its wastewater system was almost complete, including excellent wastewater tertiary treatment plants with ultra violet disinfection, but the urban streams and the nearby river and beaches were grossly polluted. The company was overstaffed by a factor of two. A Restructuring Commission was created to transform this municipal service into a modern water company, with very fast results: water losses were halved in eight months and beaches attained the European bathing water quality standards six months later. The most interesting points about this case include the initial political commitment at the top political level, outsourcing of change management process leadership, fragmentation of the existing hierarchical structure without affecting normal water and wastewater services and, from a more technical perspective, the integrated management of the urban water cycle and the entire process of NRW control.

The second case study refers to a 50 year-old water and wastewater company with 120,000 clients that was profitable, but very inefficient with 47% of NRW and providing a very incomplete wastewater service, with no treatment plant at all; only about 10% of the water supply clients were connected to the sewerage system.

The third case study refers to COPASA, in the Minas Gerais State of Brazil, one of the largest water companies in the world and which started a change management process in 2003. The company has evolved from negative results and being unable to invest for many years, to becoming a reference company in Brazil and abroad, floated on the Stock Exchange. This was a result of efforts such as medium and long-term strategic planning, good governance, ongoing employee training and other actions initiated by the *"shock management"* program implemented by the Minas Gerais State Government.

The fourth case study, also from Brazil, does not refer to a single water utility, instead it applies to virtually all water utilities in this large country. It is about the PNQS – National Prize for Quality in Sanitation-devised and managed by ABES – the Brazilian Association of Sanitary Engineering. The PNQS is much more than just a prize as it is also a very efficient institutional building tool that has been improving the quality of the workforce and management of water companies since 1997. Thousands of employees and hundreds of managers of the most important Brazilian water utilities have participated in training courses and in various projects that are part of a process geared towards quality and excellence.

The fifth case study is about change management in an unforeseen difficult context. It refers to the initial stages of change management at *Águas de Portugal*, a very large public conglomerate of around seventy water, wastewater and solid waste utilities, operating at the time mostly in Portugal and in the Portuguese speaking countries: Brazil, Angola, Mozambique and Cape Verde.

# Chapter 5.1

# Case study 1 - *Águas do Porto* (2006-2013)

*Joaquim Poças Martins*
*Coordinator of the Restructuring Commission and change management consultant*

## 5.1.1 INTRODUCTION

Porto is one of the largest and oldest cities of Portugal with the history of its water supply system able to be traced back to the 16th century. The current public water supply and wastewater systems, with 153,000 clients in 2013, have been developed over the past 120 years and were run, until 2006, by a municipal service, SMAS - Municipal Water and Wastewater Services of Porto (Aguas do Porto, 2008, 2009, 2010, 2011, 2012 and 2013).

In 2006, the Mayor decided that change and improvement were required due to the negative financial situation and public image: annual losses, excessive complaints from customers about service quality, high levels of water loss at 54% of the system input volume and polluted beaches. In particular, there was particular discomfort at the fact that beaches still failed to meet required bathing water standards even after the wastewater treatment plants of the city had been in operation for several years.

At the end of 2006 SMAS as transformed into a municipal company, *Águas do Porto* (AP), with additional responsibilities, including management of the stormwater system and rehabilitation of urban streams and beaches. Therefore, the urban water cycle, as a whole, was integrated under a single unified management structure. The organizational chart of the water utility was made very simple and flat, with only one managerial level below the Board, unlike the previous one, which was complex and with several levels of decision. A Structuring Commission reporting directly to the Board of Directors was created to develop the technical and organizational tasks that were required to transform the former municipal service into the new water company. The leadership of this Structuring Commission was contracted externally.

The change management process was geared towards the following goals, meeting the political priorities that led to the water company creation:

- To supply potable water, continuously, everywhere in the city;
- To drastically reduce water loss;
- To connect all buildings to the public sewerage system;
- To clean and rehabilitate the urban streams;

- To rehabilitate beaches to Blue Flag status;
- To manage and expand the stormwater system.

These objectives were met surprisingly soon. In just eight months the non-revenue water was halved, principally due to fast detection and repair of bursts, introduction of active leakage control and pressure management.

A new gravity-fed method of operations was developed for the water distribution system, eliminating the existing pumping stations, making the system more reliable and cost-effective to operate. This was achieved through an hydraulic analysis of the trunk mains and water tanks, resulting in significant savings in terms of energy and labour.

Within a year, pollution of the main urban streams decreased by a factor of ten. This was achieved through the elimination of sewage discharges to the stormwater system, mostly due to illegal connections from houses and industries that should have been connected to the sewerage system.

Also, in less than a year the water quality of the city's beaches met the European bathing water standards. This surprisingly fast reduction of the microbiological pollution was due to the elimination of discharges of polluted stormwater in the beaches in dry weather: this polluted stormwater was by-passed to the wastewater system for joint treatment.

These early results were achieved essentially with management and technical measures that required a thorough effort of analysis of the problems in the field and quick decisions but very little investment. Indeed, it would not be possible to prepare and realize significant investments within a few months following the mandatory public tendering procedures.

In terms of economic results, the water company moved to producing a moderate profit after significant losses over the two previous years.

Since 2006 there has been a great change in terms of customer relations. The creation of a central control room, in operation 24 hours a day, has allowed for an immediate answering of telephone calls for reporting bursts, complaining about lack of pressure or requiring specific services. All complaints are classified, recorded and displayed on the company website until they are resolved. In this way, a simple, transparent and powerful control tool was developed in order to steer and monitor the most critical aspects that affected the customer-company relationship.

The major results of the process of change are summarized in Table 5.1.1.

In 2011 *Águas do Porto* was awarded the First National Prize in the category of Reduction of Internal Costs for public companies of any sector and size in Portugal (Deloitte *et al.*, 2010). In 2011 it was also awarded an EPSA – European Public Sector Award-prize nomination for its project "Porto, a Water Sensitive City" by EIPA – European Institute of Public Administration – the leading centre of European Union learning and development for the public sector (EIPA, 2011).

## 5.1.2 COMPANY BACKGROUND AND PROFILE

The origins of the public water supply to the city of Porto can be traced back to 1597 when the first works were paid for through special taxes, namely on salt and wine. The system subsequently developed and expanded and, as in many other growing cities, water was scarce and significant urban pollution problems existed (Mesquita Arq. M. (2013), Personal Communication). Public money was even scarcer than water at the time and it was decided to proceed with the expansion of the water systems through a private concession. After a number of failed attempts after 1855, the concession of the water supply system was granted in 1880 to the French *Compagnie Générale des Eaux pour l'Étranger.* Slightly later,

**Table 5.1.1** Aguas do Porto 2006-2013.

| | October 2006 | December 2013 |
|---|---|---|
| System Input Volume ($m^3$/day) | 104,000 | 59,984 |
| Revenue Water ($m^3$/day) | 48,000 | 45,602 |
| Non-Revenue Water ($m^3$/day) | 56,000 | 13,382 (corresponding to savings of 30% of annual turnover) |
| Functions | Water Supply + Sewerage + Wastewater Treatment | Water + Sewerage+ Wastewater Treatment + Stormwater + Stream Rehabilitation + Beaches (Full Urban Water Cycle) |
| Water and Wastewater Tariff for an average family consuming 10 $m^3$/month (2006 prices) | € 1.45/$m^3$ | € 1.41/$m^3$ (Reduction of 2.8%) |
| Number of Workers | 574 | 475 (–17%) |
| Net Results | – € 3.5 million (Loss) | + € 4.0 million (Profit) |

in 1884, construction of the city's first wastewater system was contracted to the British firm Hughes & Lancaster.

The concession contract was terminated with municipalization of services in 1927, when SMAS – Municipal Services of Water and Sanitation – was created. In 1996, abstraction and the water treatment plant of SMAS were partially discontinued and integrated into a new regional bulk water system, created to serve the Greater Porto Area. SMAS was responsible for the construction and operation of the modern water supply and wastewater systems of the city until the end of 2006, when they were transformed into a new municipal water company, *Águas do Porto*.

This case study refers mainly to the restructuring of services inherited from SMAS, under the new framework of the municipal water company *Águas do Porto*, 100% owned by the municipality and publicly managed. This new water company was given responsibility for managing the urban water cycle in the city, with the exception of water abstraction and treatment, which was undertaken by a regional bulk water supply company.

The vision of *Águas do Porto* was to become a reference water company at national and international levels. Its mission was to ensure continuous service of public drinking water and sanitation throughout the city of Porto, with economic sustainability, efficiency, diligence and a sense of social responsibility, and their objectives, also presented in Figure 5.1.1 are related to the various components of the urban water cycle.

The values selected to inspire company activities were: excellence, social responsibility, transparency, trust, innovation and sustainability (Figure 5.1.2). The key indicators of *Águas do Porto* in 2013 are presented in Figure 5.1.3.

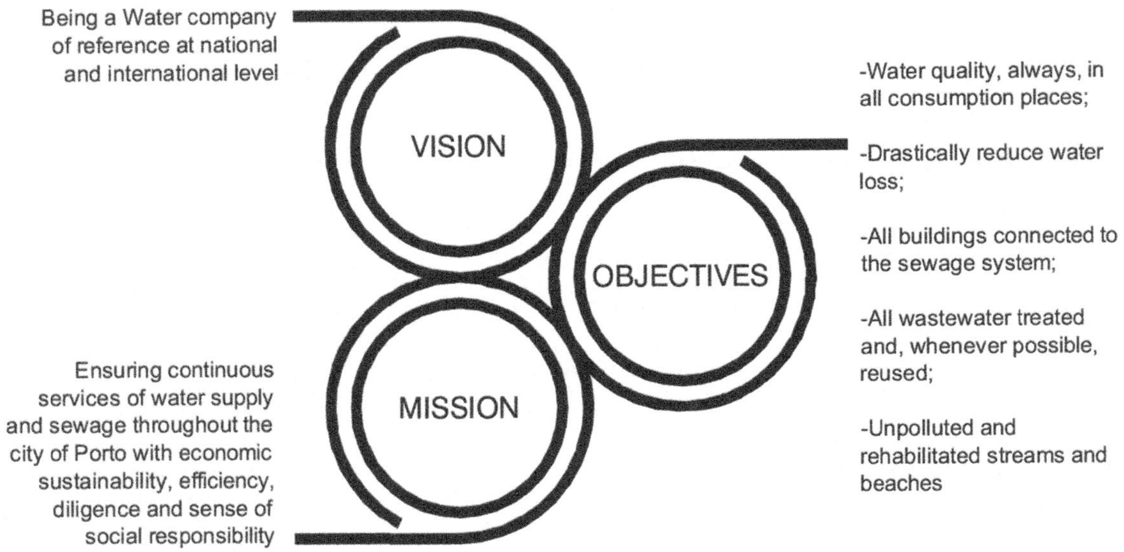


**Figure 5.1.1** Vision, mission and objectives.

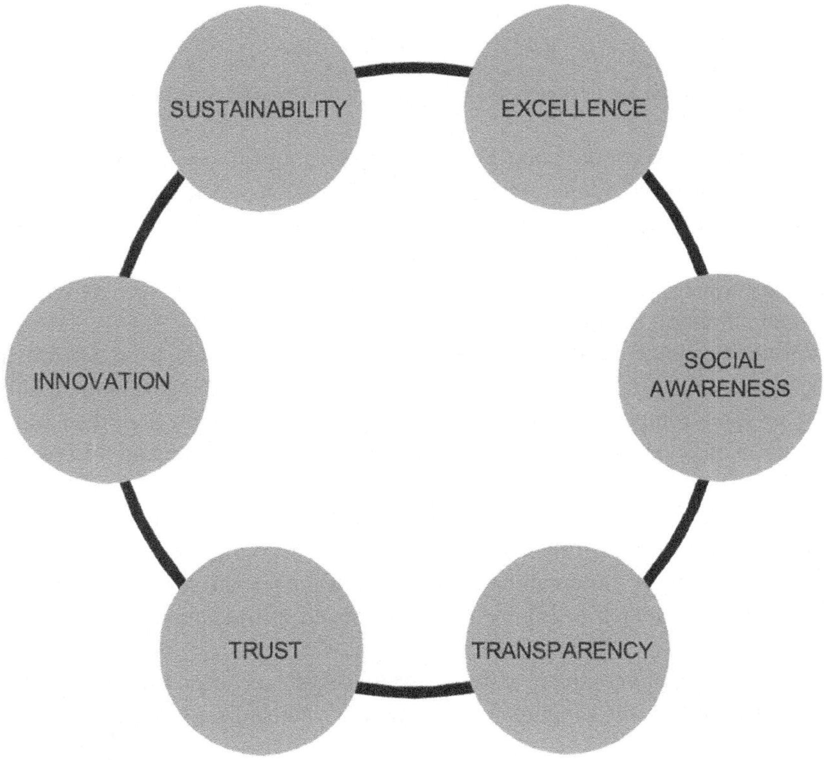


**Figure 5.1.2** Corporate values of *Águas do Porto.*

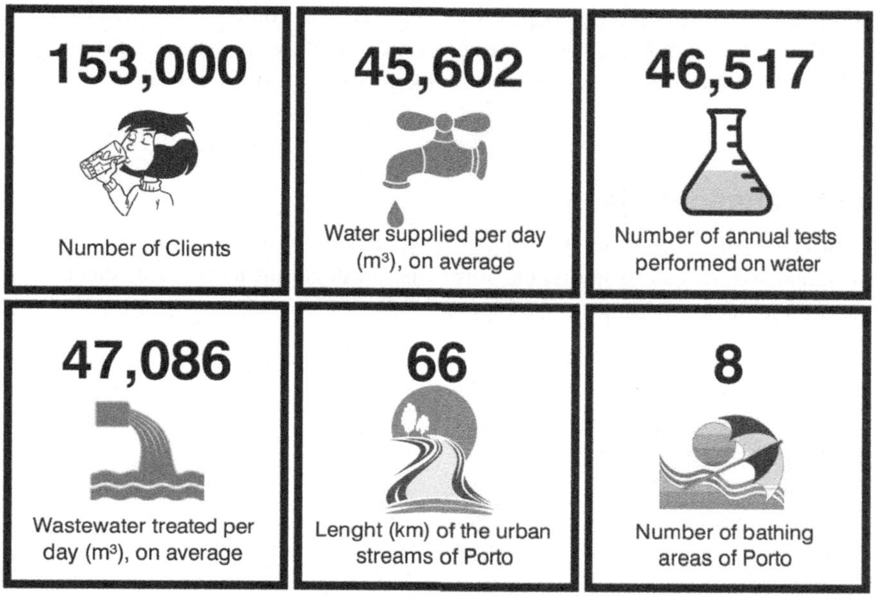


**Figure 5.1.3** Key indicators of *Águas do Porto* in December 2013.

## 5.1.3 BASIS OF THE CHANGE MANAGEMENT PROCESS

This change management process essentially followed the eight steps proposed by Kotter and was developed, most of the time, under near ideal conditions: with political support, leadership, sustainable tariffs, sufficient time and a positive context.

In this case, change meant questioning initially every aspect of the organization, appointing new managers and redefining new routines, aligned with the following objectives:

- Drastically reduce water loss from the initial level of 54%;
- Connect all remaining city buildings to the public wastewater system;
- Eliminate wastewater connections to the stormwater system;
- Rehabilitate urban streams;
- Reach "good" bathing water quality status and later obtain Blue Flag accreditation for all urban beaches;
- Move from loss to moderate profit.

### 5.1.3.1 The Restructuring Commission

A distinctive feature of this change management process was the creation of a Restructuring Commission, reporting directly to the Board of Directors, to develop the technical and organizational tasks that were required to transform the former SMAS into the new water company. The leadership of this Restructuring Commission was contracted externally.

The Restructuring Commission was led by an external consultant, included an executive member of the Board and around twelve coordinators who represented all the areas of the company. It operated in a flexible way with general meetings typically held once a week and as many steering meetings as required.

The Restructuring Commission defined around 30 reform projects, each one with a leader, clear objectives and a deadline. Some were standalone projects while others overlapped with routine activities of the water company. For example, the project to create a central control room was to be accomplished in 100 days but others, such as completion of the sewerage system, were scheduled for conclusion within three years and involved participation of many collaborators from different areas of the company and external contractors.

### 5.1.3.2 The Strategic Projects

In addition to the "normal" activity of the water utility, which consisted essentially of the operation of the water supply and wastewater systems, the following strategic projects were selected, reflecting the priorities of the change management process (Figure 5.1.4):

- Porto without Water Losses
- Porto without Pumping
- Porto with 100% Sewerage Coverage
- Porto new Stormwater System
- Porto Urban Streams
- Porto Blue Flag Beaches

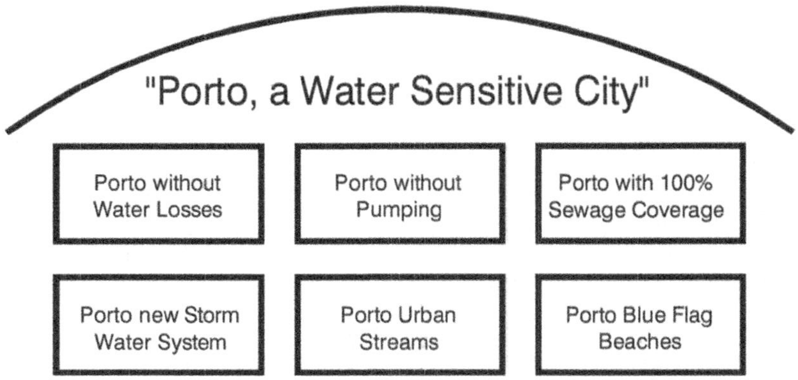


**Figure 5.1.4** Strategic projects of *Águas do Porto* in 2013.

For internal and external communication purposes, these projects have been grouped under the designation "Porto, a Water Sensitive City". This communicational aspect was given significant importance as it is certainly not enough to simply do the right things well; it is equally important to explain to stakeholders what is happening and why. It was interesting, for instance, to receive a number of positive comments from local residents saying that it was not until they saw the "Porto Blue Flag Beaches Project" on signs that identified sewerage construction sites, that they realized the link between the connection of their homes to the wastewater system and the bathing water quality in the city beaches.

Designation of projects aimed to be self-explanatory and incorporated ambitious qualitative or quantitative goals that could be associated with a vision of excellence. The Porto without Water Losses project referred to the never-ending reduction of each of the NRW components in the water balance. The Porto without Pumping project was related to a number of activities that led to the elimination of all but one distribution water pumping stations in the city, replacing them with gravity-fed flow solutions.

The Porto with 100% Sewerage Coverage project included two different objectives: namely construction of missing sewers and house connections by the water company and just as important, persuading people to give up their current connections to the stormwater system.

The Porto new Stormwater System referred to a different approach to dealing with rainwater in the city. In the past, more and more areas had been made impervious and the objective was to drain them all as fast as possible to the nearby urban streams. The increased flows could not be conveyed by the existing pipes, often leading to flooded streets and, sometimes, also to flooded basements. Instead of the traditional expensive and disruptive solution of replacing the pipes with larger diameter ones, which implys a high level of detrimental impacts on normal street life and functionality, a transition was made to the progressive selection of other types of solutions. These involved porous pavement solutions, construction of storage basins and incentives for local storage and use of rain water.

The Porto Urban Streams project was a package of three actions: eliminating polluting discharges to urban streams by mainly transferring them to the sewerage system, reopening covered river reaches whenever possible and rehabilitating bank sides. The idea was to return urban streams to the population and the wildlife, with the creation of green areas, leisure zones, footpaths and cycle ways.

Finally, the Porto Blue Flag Beaches project was related to the transformation of three kilometers of heavily polluted city beaches in 2006 into excellent nearby Blue Flag beaches, with obvious advantages in environmental, economic and social terms: water quality and biodiversity drasticly improved, residents and tourists could get to excellent free beaches in just a few minutes and local businesses flourished.

### 5.1.3.3 The first 100 days

During the first 100 days of the new water company it was possible to define the strategy, to implement a new organization chart and produce significant early results, namely:

- Eliminating virtually all visible leaks;
- Reduction of the System Input Volume by 14%;
- Replacing the main water supply pumping station of the distribution system with gravity flow;
- Creation of a Central Control Room, operating 24 hours a day;
- Receiving the first real time telemetry information regarding water flows entering the distribution zones;
- Displaying the first version of the Anomalies Chart in real time in the Central Control Room and on the company website;
- Repairing bursts on the same day that they were detected;
- Reestablishing most water supply interruptions within two hours;
- Launch of the first phase of the integrated wastewater and stormwater management which led to the first Blue Flag beaches ever in Porto.

The measures which were taken in the first 100 days had a decisive impact on the future of the organization, namely on reduction of NRW that took place during the first year. Most of these early tangible results deserved a wide attention from the media and more importantly, from customers.

## 5.1.4 THE NEW ORGANIZATIONAL MODEL

In the first week of the newly created water company, the previous organizational structure, a typical pyramidal model with many levels of decision, was changed to a much simpler and flatter one (Figure 5.1.5), with a number of advantages: faster implementation of decisions due to a reduced number of interfaces, fewer coordinating people and fewer job descriptions to define.

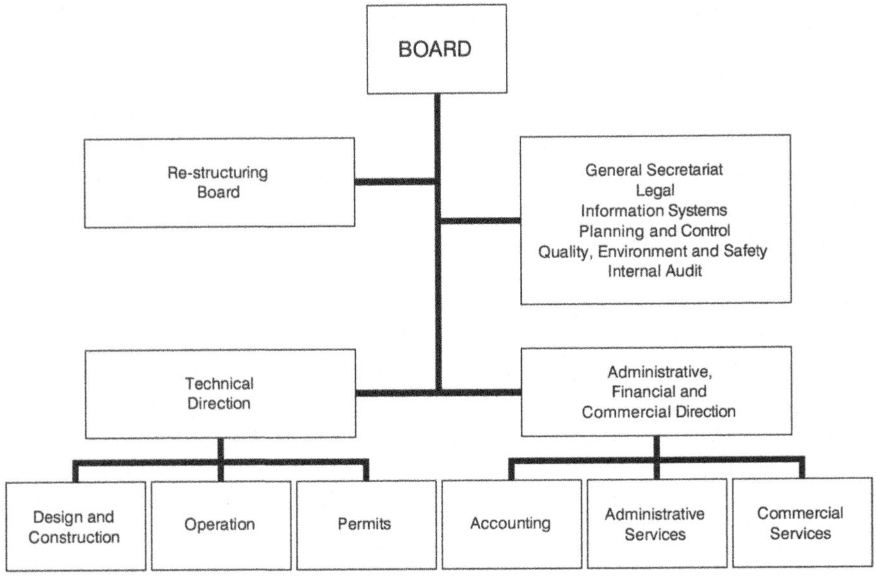


**Figure 5.1.5** *Águas do Porto* new organization chart.

Perhaps the most distinctive feature of the new organizational chart was the Restructuring Commission, reporting to the Board, with extended powers to implement the process of change over a three-year period.

Although the company was highly overstaffed, it was not socially desirable nor legally possibly to reduce the workforce number, therefore it was decided to expand areas of activity and responsibility to also include the operation of the stormwater system and rehabilitation of streams and beaches.

## 5.1.5 CENTRAL CONTROL ROOM

Surprisingly, at the end of 2006 it was very difficult to establish contact with the water utility. There were multiple operational centres, with different phone numbers and different procedures, making it difficult to effectively and efficiently regulate and control work orders and customer service requests and complaints.

The call-centre worked from 9 am to 5 pm, but it essentially redirected incoming phone calls to the different departments, without registering them or terminating them with substantive answers. Out of this period, the security officers, who controlled the entrance of the headquarters, redirected the main emergency calls to the operational teams.

One of the first decisions taken in the new water company was to create a Central Control Room, working 24 hours a day, every day of the year.

The Central Control Room (Figure 5.1.6), had typical call centre operators to receive, register and answer phone calls, specialized operators to monitor the water and wastewater systems in real time, and technical operators to interact with the teams that were in the field repairing bursts and resolving other problems.

The Control Room had three special telephone numbers: a green, a yellow and a red one. The first was to be used by company employees to report any problem they detected on the streets, the second by directors

**Figure 5.1.6** View of the Central Control Room.

and coordinators and the third was only available to the Mayor and to Board members. Obviously the red one was expected to be always available and answered immediately.

This concentration of relevant operational information in one place was one of the greatest achievements in the early stages of the new water company; it was essential to increase its global effectiveness and played a major role in the NRW reduction process.

## 5.1.6 MONITORING AND TELEMETRY

Even by the end of 2006 there was still no real-time monitoring of the networks of *Águas do Porto*. The first step to change this situation was designing a simple and expandable system to monitor water flows entering the system and levels of the water distribution reservoirs.

For reasons of robustness, flexibility, cost and speed of installation, a GSM – Global System for Mobile communications – based system was adopted with local data loggers. Subsequently, more and more flow and pressure control points have been added.

In 2013 the water supply system was divided in 19 separate distribution zones, which in turn were divided in District Metered Areas (DMA), some of which associated to Pressure Reducing Valves (PRV). Each of these separate zones with an average of 4,000 clients was equipped with one, two or three pressure transducers with data loggers and telemetry, depending on their size and topography. In addition to these more formal closed distribution zones, there were also control points in critical pipes that were continuously monitored for flow, pressure and, in a small number of pilot cases, also for noise levels using permanent acoustic loggers. All service water reservoirs were monitored for level and outlet flows and were also connected to the telemetry system.

Definition of DMAs and the installation of PRVs were based on mathematical models developed for the network. Initially, the well-known open source EPANET application from the United States Environmental Protection Agency was used.

This early installation of a real time monitoring system was essential to the NRW reduction process and allowed for control and evaluation of night flows in the DMAs.

## 5.1.7 THE ANOMALIES AND CUSTOMERS' COMPLAINTS CHART

Following a policy of transparency, *Águas do Porto* published an Anomalies and Customers' Complaints Chart on its Internet site (www.aguasdoporto.pt). This chart was updated in real time by the Central Control Room operators and presented anomalies in the water supply and wastewater systems that affected the customers and their pending complaints (Figure 5.1.7).

| TYPE | DESCRIPTION | NUMBER | OLDEST | HOUR |
|---|---|---|---|---|
| OCCURRENCES | Clients without water | 200 | 30-05-2008 | 11:11 |
| | Clients with insufficient pressure | 0 | --- | --- |
| | Pipe Bursts | 1 | 30-05-2008 | 11:11 |
| | Service Connections Bursts | 0 | --- | --- |
| | Active Leakage Control Interventions | 10 | 30-04-2008 | 20:28 |

**Figure 5.1.7** Screen Shot of the Anomalies and Customers' Complaints Chart.

Making information about the situations that affected service quality available via the Internet in real time as well as follow-up responses to customer complaints avoided unnecessary creation of tension and work backlogs, saved valuable time and stimulated employees to do things better and faster. The sooner an anomaly was detected, the faster it could be resolved. This was especially true with mains bursts, which usually lead to service interruptions and water loss.

Whenever there was a significant burst, the input volume in the corresponding distribution zone increased and pressure dropped. These values were monitored in real time and automatically compared with historical data; they were graphically displayed on the Central Control Room and on the Internet (Figure 5.1.8). Any new value that was significantly outside the expected range automatically triggered an alarm and the situation was clearly visible on the displayed graphics.

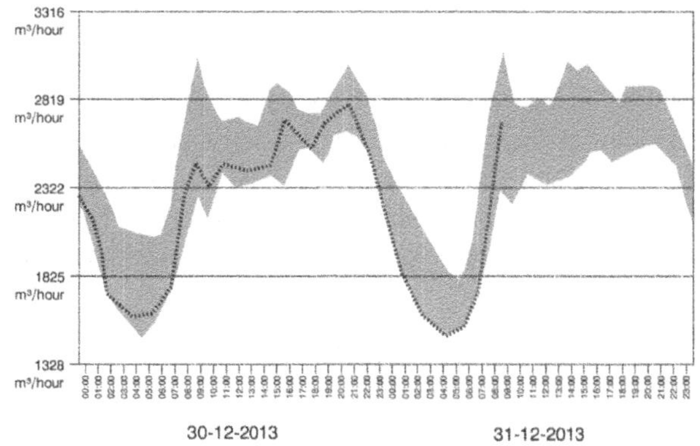


**Figure 5.1.8** Real time monitoring of the input volume into a distribution zone.

Data loggers were programmed to send information on hourly basis; it would be possible, of course, to select a shorter time interval, but this would mean much more data to process and higher communication costs without significant benefits. Of course, when there was a great burst, the Control Room was alerted by phone by affected customers or by people who happened to be passing the burst location.

## 5.1.8 THE REVENUE ANOMALIES OFFICE

At *Águas do Porto*, as in any other any water company, there was a certain amount of water that was used and not converted into revenue, related to authorized un-billed consumption, unauthorized consumption and customers under reading meter errors. There were also unpaid bills, which defeated collection attempts and were added to customer debt in the company accounts.

These consumptions related to unpaid bills are not classified as NRW in the IWA water balance, but correspond to water that is not converted into revenue. Indeed, these unpaid bills are indirectly charged to the paying customers through the tariffs and therefore must be reduced as much as possible.

A Revenue Anomalies Office (Figure 5.1.9), was specifically created to reduce customer debt and unauthorized consumption in a systematic and integrated way. It included the operators in charge of cutting off consumer water supply in cases of long-term debt and legal officers with delegated powers to prosecute any customers involved in unauthorized water consumption and to take legal action to recover debt. This office also had a strong functional connection with the meter-reading officers because debt and illegal consumption are often associated with the lack of access to customers' meters.

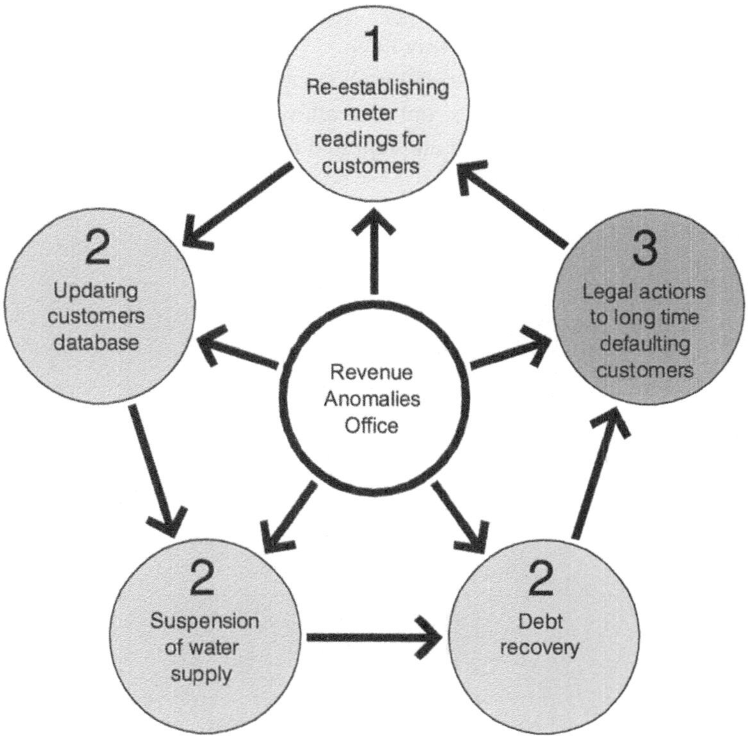


**Figure 5.1.9** Revenue Anomalies Office.

The integrated management of the teams that were out in the field to read the customers meters, to inspect private connections and to cut-off supplies in case of repeated non-payment of water bills, and the coordinated action with officers in charge of taking legal action against non-payers and customers with unauthorized consumptions was very successful.

It was possible to reach a situation in which an average of 98% of customers actually paid their bills (95% in the more problematic areas of social housing) and with a significant recovery of debt. This was achieved through a number of measures:

- Re-establishing meter readings for customers who had previously failed to provide access to the water meter, noting that over 60% of water meters were located inside houses, in private areas;
- Installation of AMR – Automated Meter Reading – in selected cases;
- Permanently updating the customer database with results of every attempt to collect debts;
- Suspension of water supply to long-term defaulting customers;
- Active detection and inspection of faulty meters;
- Active detection of illegal situations, including illegal connections and pampered meters;
- Taking legal action in case of reiterated debt and illegal situations.

The annual volume of not measured (estimated) revenue water dropped from 14% to 6% through adoption of a different meter reading strategy: meter readers were transformed into zone managers and a decision was made to install AMR in the consumption points of more difficult access. Each of the meter readers was allocated a zone and his/her work was evaluated through a specific performance indicator (N2,N4,N6,N12), the number of customers in the zone whose meters had not been read the more than 2, 4, 6 or 12 months. The ideal value of the indicator was, of course (0,0,0,0). Figures 5.1.10 and 5.1.11 show the situation evolution concerning the N6 component of the performance indicator over a short period of time, just following the creation of the Revenue Anomalies Office.

The meter readers went to the field with small mobile computers (tablets) connected to the Internet. These were used to register normal readings and also allowed the recording of any anomaly that was detected. These anomalies were photographed and reported in real time to the back office.

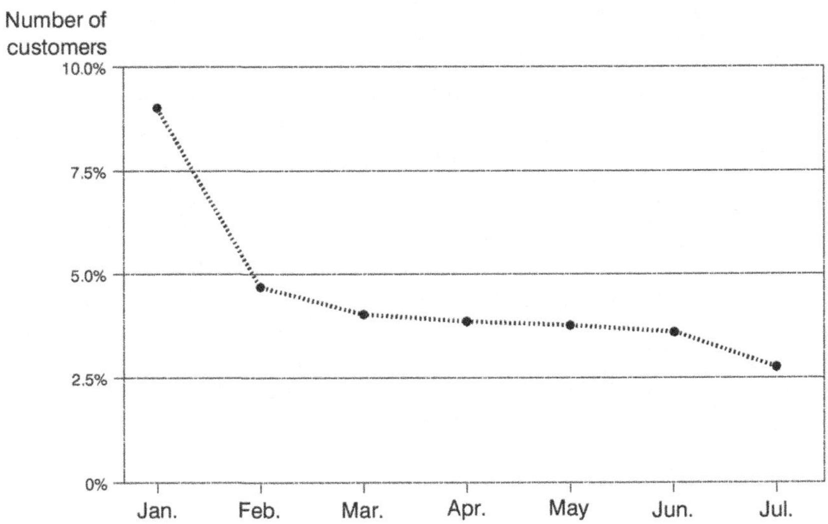


**Figure 5.1.10** Reduction of the customer meters that were not read for more than six months (total).

**Figure 5.1.11** Evolution of the customer meters not read for more than 6 months in the metering zones.

Even with all these attempts, there were always houses that were only used from time to time and others that had been abandoned without cancellation of the water supply contract; it was relatively common to detect leaks, some of which were very large, in houses that could not be visited for a very long period of time.

To help identify and reduce the number of visits to buildings with suspected revenue anomalies it was decided to combine all relevant information concerning customers in these buildings. This information was updated and made available in real time by the meter readers/zone managers who were in the field daily using the above mentioned mobile computers connected to the Internet.

Daily analysis of this comprehensive information on the customers, including GIS maps and images of the buildings, obtained directly and/or from street map views available on the Internet, played an important role in the combat against revenue anomalies and fraud. From the customer addresses, for instance, it was possible to identify which apartments were more likely to have the same typology and therefore could be expected to have similar consumption patterns. Alternatively, the external building image allowed for a better understanding of individual situations.

## 5.1.9 NRW REDUCTION PROCESS

Perhaps the most interesting aspect of this case study is the analysis of NRW evolution over time and its relation with measures that have been taken (Figure 5.1.12). The reduction was faster in the initial stages during 2007 and 2008 essentially because the original values were very high and there were many problems that could be resolved relatively fast, mainly through management measures, with very little investment, namely eliminating all visible leaks and starting the repair of bursts almost immediately after they had been reported and detected. Active Leakage Control teams were formed and sent to the seemingly more problematic distribution zones, mostly at night, including weekends.

These measures did not result in significant additional costs because human resources and equipment were already there: the cost of eliminating a leak or repairing a burst is the same whether it happens an hour, a day or a week later whilst the value of the water saved is, of course, very different. The corresponding costs in terms of the water utility image and impact on employee attitudes are also substantially different.

In 2006 the SIV – System Input Volume – was 104,000 m$^3$/day whilst RW – Revenue Water – was 48,000 m$^3$/day. Some eight months later, consumption of 76,000 m$^3$/day from the bulk supplier was already sufficient to supply the same volume of water (48,000 m$^3$/day) to the clients. NRW was halved from 56,000 to 28,000 m$^3$/day, almost entirely through the reduction of real losses (Figure 5.1.13).

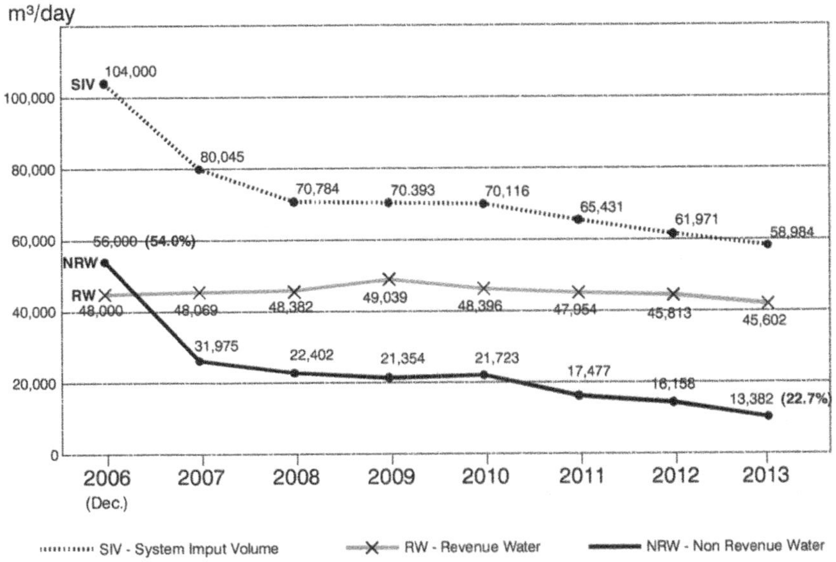


**Figure 5.1.12** Evolution of NRW, SIV and RW.

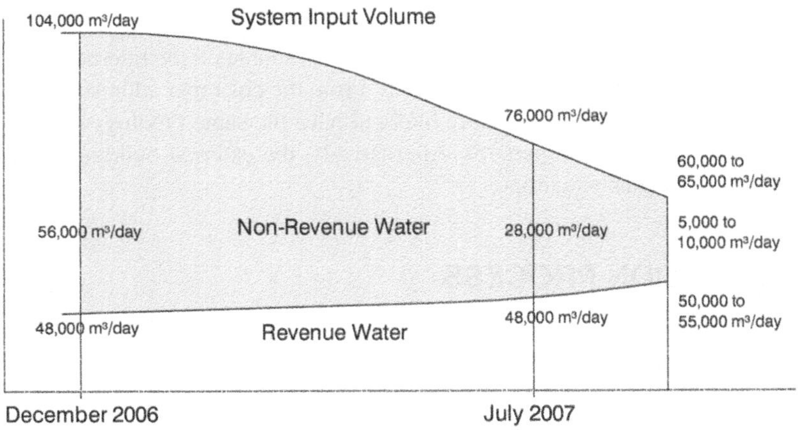


**Figure 5.1.13** Halving Non Revenue Water in eight months and early future perspectives.

At the time it was estimated that the ELL – Economic Level of Loss – should be between 5,000 and 10,000 m³/day and could be achieved through a reduction of the SIV (reduction of real losses) and an increase of the RW (more customers and reduction of apparent losses). In fact, the RW globally decreased from 2009 due to the economic crisis, in spite of the significant increase of the consumption of the customers who had their old and faulty meters replaced or who had illegal connections.

The next measure aimed to reduce pressure simply through opening and closing valves to alter distribution zone limits.

Major pipe replacement interventions to reduce NRW didn't take place until the third year after the creation of the new water company, whilst even then, initially, only short lengths of pipes were replaced in case of recurrent bursts. It was interesting to notice the impact on NRW reduction of the significant replacements that occurred during the fourth year.

Due to the reduction of public investment from 2011, related to the economic crisis, the mains replacement program was substantially delayed. However, it was possible to accelerate less capital intensive measures such as advanced pressure management projects and the repair of service water tanks; hence it is these activities which explain the reduction of NRW that continued to take place from 2011.

The reduction of NRW was due mainly to the reduction of real losses: the minimum night flow decreased substantially and so did the SIV in the distribution zones.

The whole process of reducing NRW was inspired on the well-established water balance, developed by the IWA Water Loss Task Force, and included the following main activities:

- Permanent data mining in the corporate information system, namely in the customer and assets databases;
- Local inspection of the distribution zone every time there was an unexplained increase of the night flows;
- Fast repair of operational bursts;
- Active leak detection and location using both traditional listening sticks and state of the art acoustic technology;
- Real time flow and pressure monitoring at key points in the network;
- Pressure control, through a redefinition of the distribution zones and installation of PRVs – Pressure Reducing Valves.
- Billing of all consumption and services;
- Taking legal action in case of theft or non-payment;
- Inspection or replacement of old or suspicious customers meters;
- Installation of meters with automatic reading in places of difficult access;
- Replacement of pipes with frequent bursts and high levels of background leakage.

Between 2006 and 2013, NRW was reduced by 76%, from 66,000 $m^3$/day to 13,682 $m^3$/day. During the second half of this period there was a significant reduction of customer numbers, the consumption per customer and the total consumption, which made it more difficult to interpret the values of the Revenue Water.

The gains with NRW reduction in the first year amounted to around 30% of the total annual water utility revenue. In the first seven years, the annual savings through NRW reduction corresponded in average to 14% of the annual revenue.

### 5.1.10 REPLACING PUMPING STATIONS WITH GRAVITY FEEDS

As a result of the local topography, a great part of the treated water supplied by the regional bulk water company to the Porto municipal distribution system had to be delivered to a service reservoir located on high ground outside the city at an elevation of 180 meters, whilst the maximum elevation of the city was only 160 meters (Pocas-Martins, 2007). Therefore it was incongruous that there were so many pumping stations in the water supply distribution system in the city for decades (Figure 5.1.14). The buildings and water tanks have been considered for alternative uses (Box 5.1).

**Figure 5.1.14** Pumping stations.

Through detailed analysis of the network, including mathematical modelling and taking into consideration the reduction of flows in the distribution system due to the significant water loss reduction, it was possible to simulate that it would be possible to eliminate the six existing major pumping stations and replace them almost completely with gravity flow, with very little alteration of the network, mainly opening and closing valves. The concept is illustrated in Figure 5.1.15.

Following this operational reorganization, areas of the city in which clients were supplied by pumping were very much reduced as shown in Figure 5.1.16. The only remaining water pumping station was of the variable speed type, with a much lower pressure head than previously. Figure 5.1.17, shows that the annual energy consumption for pumping was drastically reduced and approached zero.

Analysis of the network also showed that it was very much of the root and branch type. A number of bursts occurred after the new water company was established and there was no alternative to supply vast areas of the city while these bursts were being repaired. The situation improved very quickly due to two main measures: the first and most immediate was reducing pressure in many mains, leading to a reduction in burst frequency; in addition, it was possible to create more connection and redundancy within the network, creating loops of water mains through the installation of several small sections of additional mains and again, opening and closing valves.

The great reduction of water loss had a direct impact on the required storage capacity of the distribution system reservoirs, contractually set with the bulk water company at two days retention time, to take into account eventual interruptions of supply of treated water.

The System Input Volume was halved and so did the necessary storage capacity of the distribution system reservoirs in the municipal system. The existing reservoirs became more than sufficient to meet the required two days retention time and it was possible to avoid the budgeted expenditure for expansion of storage capacity and save the corresponding construction costs, equivalent to one third of the company annual turnover. Also, due to the greater inter-connectivity and redundancy of the network, the storage capacity of each reservoir could be mobilized to also serve the normal areas of influence of other reservoirs.

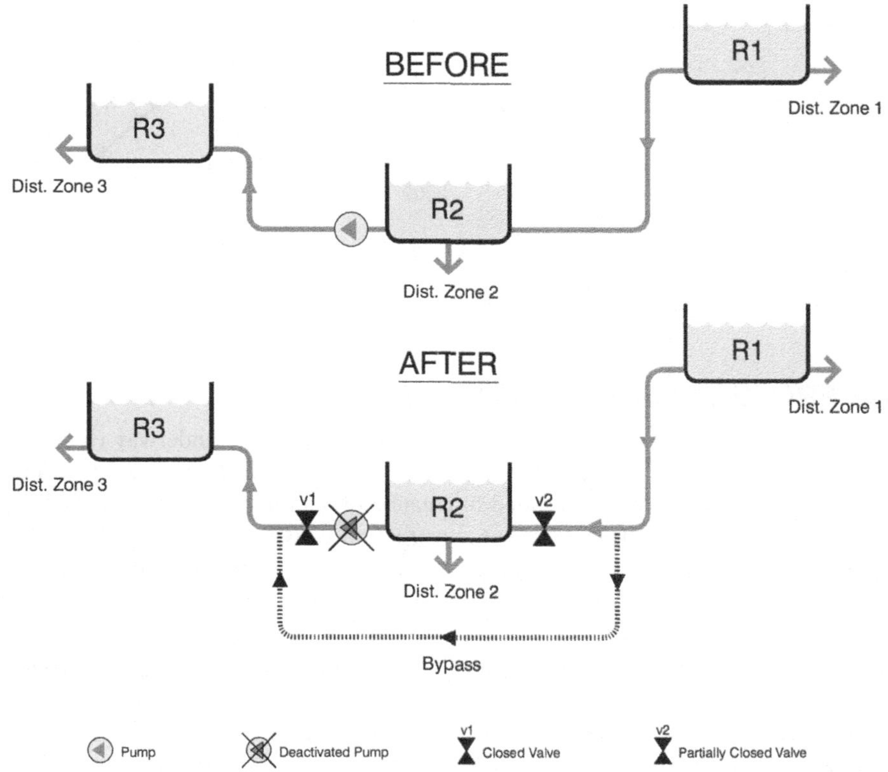


**Figure 5.1.15** Replacing pumping with gravity flow.

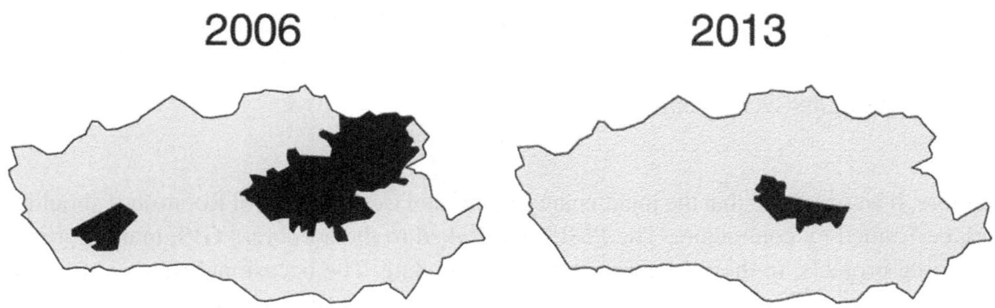


**Figure 5.1.16** Reduction of the distribution zones supplied by pumping.

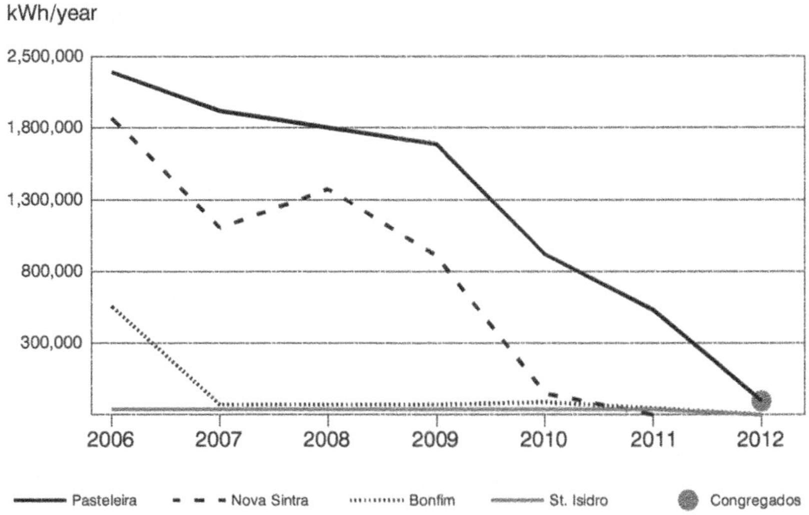


**Figure 5.1.17** Reduction of energy consumption in pumping.

## 5.1.11 THE PROJECT MANAGEMENT OFFICE

*Águas do Porto*, like most water utilities, had many construction projects underway or planned, in order to expand, rehabilitate and repair its infrastructure. To manage the large number of water supply and wastewater construction contracts it was decided to create a project management office (PMO), with the mission to contract and supervise the design and construction works, keeping to the time schedule and keeping budgets under control. Initially it was necessary to contract external project management services but at a later stage it was possible to benefit from previous experience and establish an internal PMO.

At *Águas do Porto,* project life cycles were mainly divided in three phases as shown in Figure 5.1.18. It was necessary to perform the design, procurement and construction phases simultaneously for many different types of projects with many different scopes and locations. With that in mind, a project management information system (PMIS) was implemented, which could integrate information produced during the design phase and communicate it effectively to the team members involved in the procurement and construction phases.

**Figure 5.1.18** Project Life Cycle adopted for *Águas do Porto.*

Furthermore, it was realized that the maintenance teams and Central Control Room staff should be aware of the work performed by contractors. The PMIS was linked to the corporate GIS, making it possible to display on-going projects, in their different phases, at any date. The occasional short-term construction sites for the installation of house connections or burst repairs were also included in the system, making it possible, for instance, to anticipate potential traffic conflicts and to facilitate the coordination with

**BOX 5.1 A SECOND LIFE FOR WATER TANKS**

In 2007, *Águas do Porto* organized an international architectural design competition, named "Innovative Ideas for Porto Water Tanks" (figure above), open to young architects and architecture students, for the rehabilitation of the seven water tanks in the city and its surrounding spaces.

It was essential that the space would be open to the public and that the site might support an economically viable activity that covered both construction and running costs. Examples of activities include bars, restaurants, bookshops, golf training or field sports facilities.

This contest, attracted a large national and international participation. Proposals were received from 173 teams coming from 21 countries as far away as Japan, Israel and Korea. One of the winning proposals is presented in the following figure.

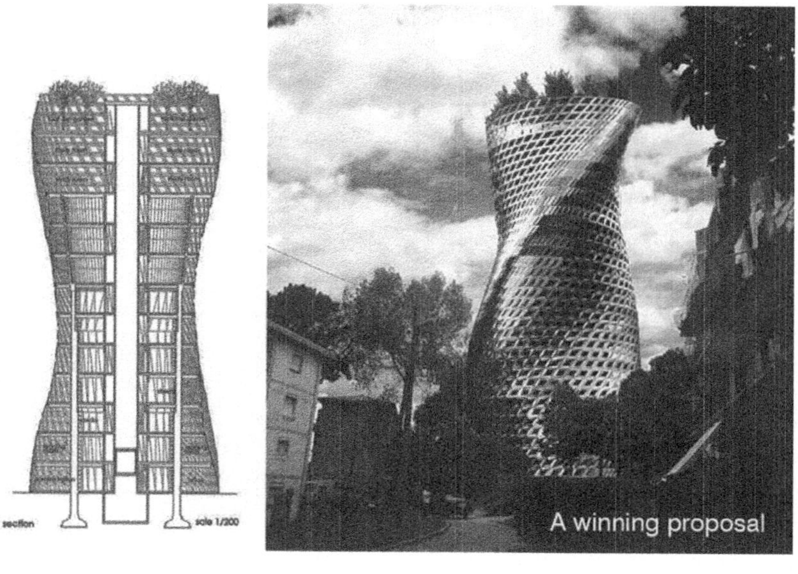

other public works going on in the city. Figure 5.1.19 shows a GIS based screen shot of the program with representation of several projects in their varying phases. The graphical representation may be used to provide user friendly access to the underlying project databases. All the existing and planned construction sites, along with their forecasted schedules, were included in the program. This allowed for eventual conflicts between works to be anticipated and avoided.

The integrated management of construction works also had a very significant impact on contract duration and respective costs. The average length of pipes installed per day increased by 50% (Figure 5.1.20). The unit costs corresponding to the construction of new water supply and sewers decreased by 50% from the creation of the water company (Figure 5.1.21), although part of this reduction is also due to

**Figure 5.1.19** GIS screen shot with live global information on the projects.

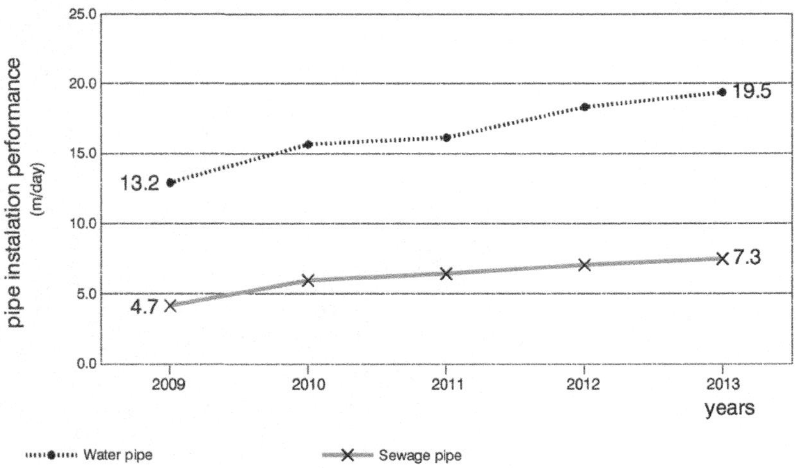


**Figure 5.1.20** Pipe installation performance from work commencement.

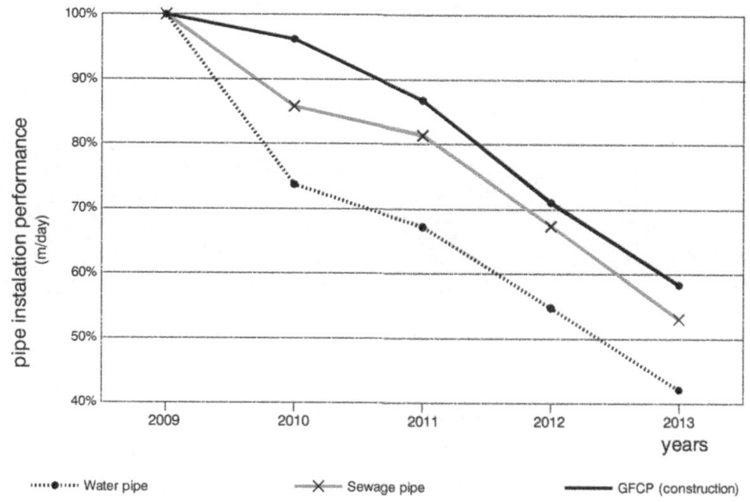


**Figure 5.1.21** Evolution of the pipe construction prices.

the market conditions related to the economic crisis, as shown by the decrease of the Construction Gross Fixed Capital Formation (GFCF).

The increase in performance indexes is directly proportional to the improvement of stakeholder management policies implemented, improved planning processes introduced and greater transparency objectives outlined for the entire project life cycle, even in the present case, where specifications were modified in order to increase the final quality of the new infrastructures. In conclusion, in addition to the cost reductions, implementation of the PMO also resulted in an improvement in the works quality and a substantial reduction of construction times, consequently lowering their impact on city life.

## 5.1.12 INTEGRATED MANAGEMENT OF WASTEWATER AND STORMWATER SYSTEMS

Even just 30 or 40 years ago, the sewerage system of Porto only covered a small part of the city and the municipality issued new building permits assuming that these would have septic tanks connected to the stormwater system, which in practice, was transformed into a combined system.

Since then, a decision was made to adopt separate systems for wastewater and stormwater and in 2006 the sewerage system was close to completion. However, a very difficult task was still ahead: to locate properties connected to the stormwater system and, even more challenging, to identify properties that were connected to the sewerage system but still had partial connections to the stormwater system. Because of these missing or illegal connections, there were clear signs of microbiological pollution in the stormwater system. The polluted stormwater was discharged into the urban streams and directly or indirectly, on to the beaches.

To protect the beaches, it was decided to install an interceptor sewer close to the beaches (Figure 5.1.22), which would receive discharges of polluted stormwater that had not been provisionally diverted to the wastewater system upstream. This polluted water was then pumped from the interceptor sewer to nearby sewers and, therefore, to the wastewater treatment plant.

**Figure 5.1.22** Construction of the interceptor sewer for wastewater and rainwater.

## 5.1.13 REHABILITATION OF URBAN STREAMS

Urban streams in Porto, like in so many other cities, were diverted from their original watercourses, confined into concrete walls and pipes, buried below streets and buildings and polluted by domestic and industrial discharges. The new municipal company *Águas do Porto* also became responsible for management of these urban streams. Cleaning, re-opening and rehabilitating were the three main objectives of the project, which aimed to establish a compromise between the urban environment and natural resources of the city, allowing ecosystem recovery and improving their surroundings for use by citizens. Cleaning streams required buildings to be properly connected to the municipal sewerage network and sewage to be transported to Wastewater Treatment Plants.

Where possible, streams were re-discovered and re-opened with demolition of previously enclosed sections (Figure 5.1.23), with the following benefits:

- Reduction of flooding risks;
- Easier detection of pollution sources and illegal discharges;
- Improvement of water self-purifying capacity;
- Increase of ecological connectivity (longer interlinked green areas);
- Creation of leisure areas and more opportunities for contact between the local population and nature.

**Figure 5.1.23** Demolition work to reopen urban streams.

Interventions along the stream bank sides aimed to achieve various parameters of aesthetic quality and ecological continuity through adoption of native species. Possible solutions depend on the space available and as Figure 5.1.24 shows, contribute to make the city greener, with more footpaths and cycle ways. Rehabilitated rivers and streams enhanced new mobility solutions that favour movement of pedestrians

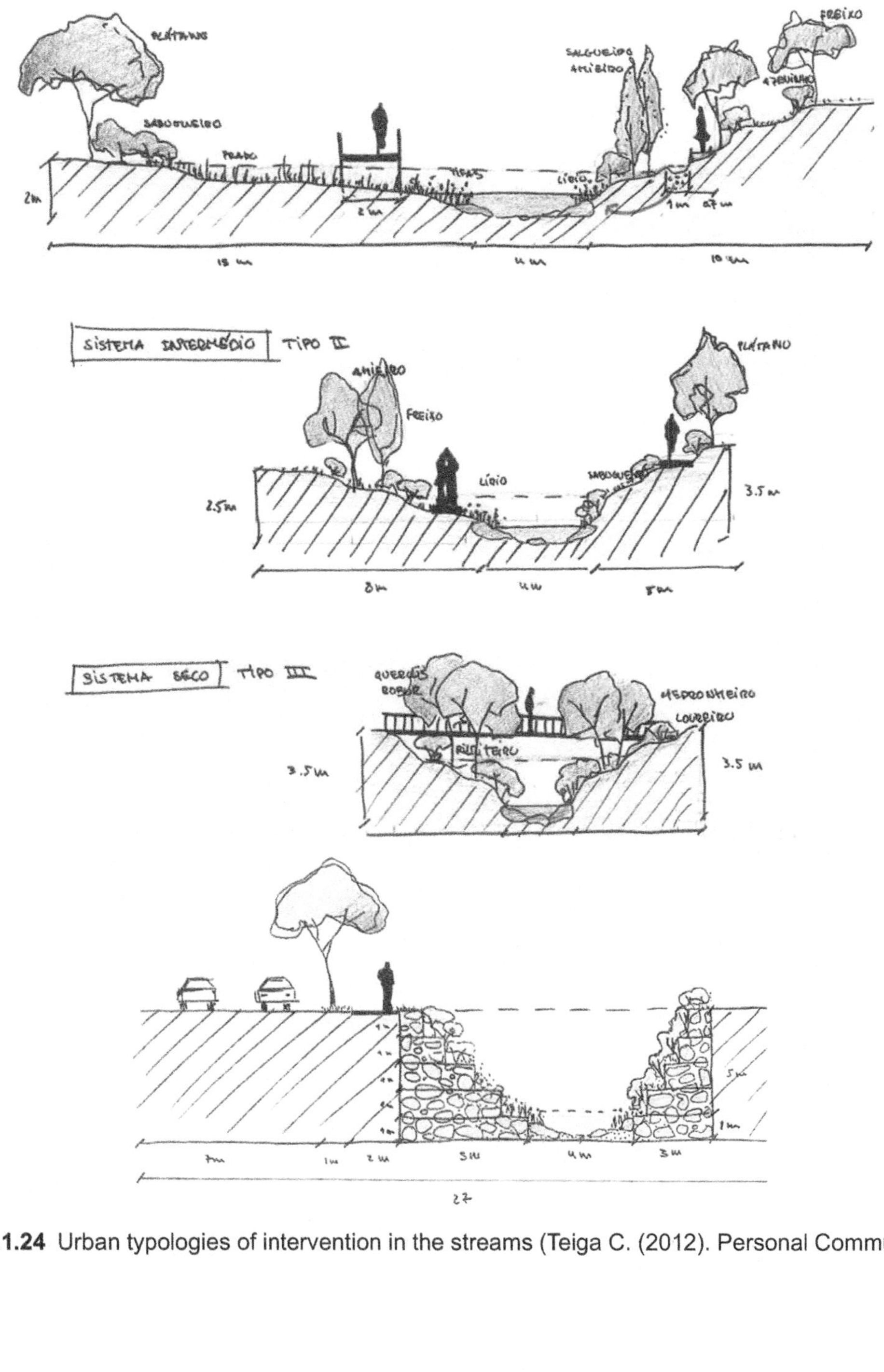


**Figure 5.1.24** Urban typologies of intervention in the streams (Teiga C. (2012). Personal Communication).

and cyclists, contributing to the overall sustainability of the city. Figure 5.1.25 shows examples of environmentally sensitive rehabilitation projects of urban streams in the city.

However, when an urban stream is enclosed in a covered channel, there is a risk that a collapse such as the one presented in Figure 5.1.26 may occur. In some cases of dense urban occupation, the only practical solution was to repair the walls, floor and ceiling of the covered stream or replace the affected reach with an enclosed box. However, whenever possible, it was certainly a much better option to re-open the stream and integrate it into the wider city environment.

**Figure 5.1.25** Examples of rehabilitation of an open urban stream.

## 5.1.14 BLUE FLAG BEACHES

In 2006, two years after the commissioning of the second and last of the city's wastewater treatment plants, the water quality in the beaches continued to be insufficient to meet European bathing water standards. At the time, it was accepted that pollution of the beaches was due to the nearby River Douro, which received significant untreated discharges from a number of municipalities along the course of the river. This, in some part, explains the water quality of the beaches, which had remained bad for years.

**Figure 5.1.26** Examples of rehabilitation of a covered urban stream.

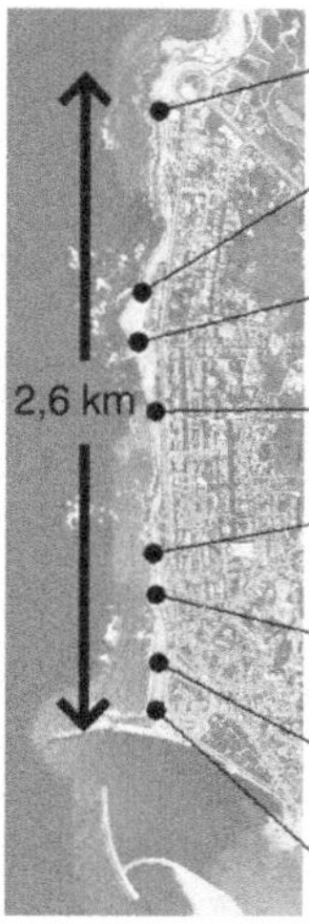


| Beach | 2005 | 2006 | 2007 | 2008 | 2009 | 2010 | 2011 | 2012 | 2013 |
|---|---|---|---|---|---|---|---|---|---|
| Castelo do Queijo | × | × | ✓ | ✓ | ✓ | ✓ | ✓ | ✓ | ✓ |
| Homem do Leme | - | - | ✓ | Blue Flag | Blue Flag | Blue Flag | Blue Flag | Blue Flag | Blue Flag |
| Molhe | - | - | ✓ | ✓ | ✓ | Blue Flag | Blue Flag | Blue Flag | Blue Flag |
| Gondarém | × | × | ✓ | ✓ | ✓ | Blue Flag | Blue Flag | Blue Flag | Blue Flag |
| Luz | - | - | ✓ | ✓ | Blue Flag | Blue Flag | Blue Flag | Blue Flag | Blue Flag |
| Ingleses | - | - | ✓ | ✓ | Blue Flag | Blue Flag | Blue Flag | Blue Flag | Blue Flag |
| Ourigo | - | - | ✓ | ✓ | Blue Flag | Blue Flag | Blue Flag | Blue Flag | Blue Flag |
| Carneiro | - | - | - | - | - | Blue Flag | Blue Flag | Blue Flag | Blue Flag |

- Unclassified area (without information); × Not meeting the Bathing Water Standards; ✓ Meeting the Bathing Water Standards; Blue Flag

**Figure 5.1.27** Evolution of the bathing water quality in the urban beaches of the city.

Figure 5.1.28 Improved biodiversity on the Porto coastline.

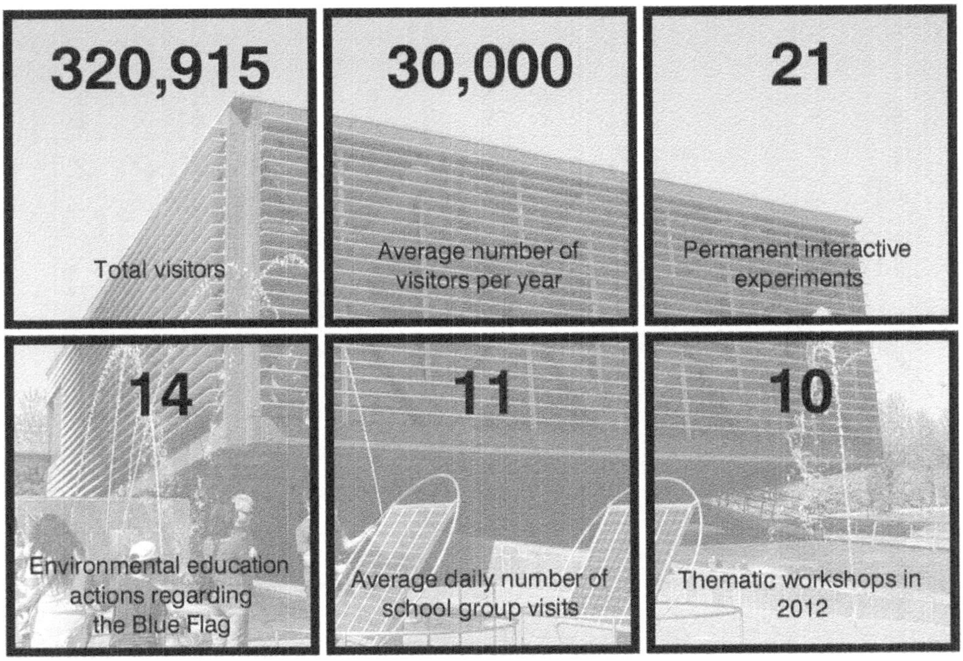


Figure 5.1.29 Environmental education: the Water Pavilion.

However, from past experience and through analysis of available data, it was apparent that the major reason for pollution of the beaches was not the nearby rivers but indeed the discharges of polluted stormwater either directly on the beaches or indirectly through the urban streams. Indeed, even in dry weather there was water on the stormwater pipes that was being discharged on the beaches with analysis results showing that this water was heavily polluted. This was certainly due to illegal sewage connections because, although the sewerage network was almost complete, there was still a significant number of missing house connections in the basin that drained to the coast.

It would take a few years to connect all of them to the wastewater system; therefore, it was decided to intercept the major polluted stormwater discharges and to connect them to the wastewater system during the summer months (dry weather collection). This was done in just three months and the bacteriological pollution on the beaches decreased by a factor of about thirty, meeting the European bathing water standards.

Construction of the missing sewers and property connections was undertaken progressively over the following years, but these simple emergency measures have contributed to the cessation of the unacceptable "swimming forbidden" status of the urban beaches in the city during the first year of the company's activity.

Evolution of the classification of the beaches is presented in Figure 5.1.27. The biodiversity of the coastal zone has since started to show a significant improvement, with the return of species that had been not been present in the area for many years (Figure 5.1.28). The urban water cycle was completed with the integration of the Water Pavilion into *Águas do Porto*, (Figure 5.1.29). Formerly, the Pavilion hosted experiments and a laboratory related to the water cycle. New experiments were added, featuring water and wastewater treatment, water loss control in supply networks and efficient use of water and energy at home. The Water Pavilion received more than 30,000 visits a year, mainly groups of school children with their teachers.

## 5.1.15 REFERENCES


Aguas do Porto (2008). Annual Report 2007 (in Portuguese). Aguas do Porto, Porto, Portugal.

Aguas do Porto (2009). Annual Report 2008 (in Portuguese). Aguas do Porto, Porto, Portugal.

Aguas do Porto (2010). Annual Report 2009 (in Portuguese). Aguas do Porto, Porto, Portugal.

Aguas do Porto (2011). Annual Report 2010 (in Portuguese). Aguas do Porto, Porto, Portugal.

Aguas do Porto (2012). Annual Report 2011 (in Portuguese). Aguas do Porto, Porto, Portugal.

Aguas do Porto (2013). Annual Report 2012 (in Portuguese). Aguas do Porto, Porto, Portugal.

Deloitte, Institute for Public Administration, Luso American Foundation and *Diário Económico* (2010). Good Practices in the Public Sector: Reduction of Internal Costs. http://www.boaspraticas.com/prm_gala_vencedores.php, last accessed on 2014.05.30.

EIPA (2011). Porto, a Water Sensitive City: Sustainable Management of the Urban Water Cycle. http://www.epsa2011.eu/en/content/show/&tid=162, last accessed on 2014.05.30.

Pocas-Martins J. and Cunha R. (2007). Porto Gravity Flow Project: Reformulation of a Large Municipal Water Supply System, Jornadas de Hidráulica Recursos Hídricos e Ambiente, IHRH, Faculty of Engineering, University of Porto, Porto, Portugal.

# Chapter 5.2

# Case study 2 - *Águas de Gaia* (1998-2004)

*Joaquim Poças Martins*
*CEO (1998-2004)*

## 5.2.1 INTRODUCTION

Vila Nova de Gaia, centre of the world famous Port wine trade, with 300,000 inhabitants, is the third largest municipality of Portugal in terms of the resident population. Until 1998, the Municipal Services of Gaia (SMG) managed to provide a reasonably effective water supply service, but with a very high levels of NRW of around 47%. The wastewater service was rudimentary, with only 10% of properties connected to the public sewerage system. There were no wastewater treatment plants at the time; therefore, raw sewage was discharged directly into polluted urban streams, the River Douro and the sea. Consequently, the popular and well-used sandy beaches along the 17 km coastline were grossly polluted, failing to meet the European bathing water standards by a significant margin.

The Municipal Services had 300 employees for its 80,000 water supply and 15,000 wastewater customers, out of 120,000 potential customers. There was an inherent lack of working conditions in terms of specialized equipment, vehicles, computers, communication systems, tools, changing rooms and uniforms. In addition, a complex, heavy, hierarchical and bureaucratic organizational structure was in place, with too many levels of decision, typical of a traditional public service.

In 1997, the newly appointed mayor realized that the existing municipal services were not adequate to serve the population and took the political decision to prioritize its restructuring, making it clear that the whole process had to be financed without increasing the municipal budget or customer tariffs. To manage this change process, the mayor appointed a new professional Board of Directors and decided on the transformation of the municipal services into a municipal water company, with the name *Águas de Gaia*.

With clear political support, professional management, motivated workers and the possibility to apply for financial support from European Union structural funds, it was possible to initiate a change management process with the following priority objectives:

- To complete the water supply system in order to guarantee potable water of high quality, 24 hours per day, everywhere in the municipality;
- To complete and extend the wastewater system and to connect all properties in the municipality to the public sewerage system;
- To clean and rehabilitate the urban streams;

- To rehabilitate and obtain Blue Flag status for all beaches in the municipality along17 km of coastline;
- To pass to the customers and the general public a new image as a modern water utility and to provide a high service quality.

The results of this process are summarized in Table 5.2.1.

**Table 5.2.1** *Águas de Gaia*, Portugal, 1998-2004.

| | 1998 | 2004 |
|---|---|---|
| Functions | Water Supply + Sewerage (No Wastewater Treatment Plants) | Water + Sewerage+ Wastewater Treatment+ Stormwater + Stream Rehabilitation+ Beaches (Full Urban Water Cycle) |
| Water Customers | 80,000 | 120,000 |
| Wastewater Customers | 10,000 | 110,000 |
| Wastewater Treatment Plants | 0 (0%) | 5 (100%) |
| Investment 1998 – 2004 (1998 Base Costs) | 150 million Euros | |
| Average Domestic Water Supply Tariff (1998 Base Costs) | 0.75 €/ $m^3$ | 0.73 €/ $m^3$ |
| Production/Acquisition Cost of Treated Water (1998 Base Costs) | 0.10 €/ $m^3$ | 0.28 €/ $m^3$ |
| Non-Revenue Water | 28,085 $m^3$/day (47% of SIV) | 11,024 $m^3$/day (20% of SIV) |
| Number of Workers | 300 | 300 |

As a result of this change management process, NRW was reduced from 47% to 20% over six years. The corresponding savings helped to finance, without tariff increases, a new wastewater system built almost from scratch: 700 km of new sewers, five wastewater treatment plants and 100,000 new home connections. The principal urban streams were successfully cleaned and rehabilitated whilst the 17 km of beaches have been improved to Blue Flag status, the highest quality level in Europe. Over 30 km of footpaths were built along the urban streams and coastline, allowing people to benefit from unpolluted waterfronts (*Águas de Gaia*, 1999, 2000, 2001, 2002, 2003, 2004 and 2005).

## 5.2.2 MODERNIZING MUNICIPAL SERVICES

The municipal services were initially overstaffed, even considering the need to expand the water supply system and especially, the wastewater systems. Reducing the workforce through dismissals was not an option, therefore, expanding the range of activities was considered. Management of the stormwater system was transferred from the municipality to the water company and it was decided to initiate two new activities: rehabilitation of urban streams and beaches. This resulted in the amalgamation of all components of the urban water cycle in the same municipal organization, which proved to be very effective in terms of environmental results whilst increasing efficiency. For instance, the workforce and equipment used in wastewater system operations have been organized to operate the stormwater system as well.

For the public, the message transmitted was simple and clear: *if you have or know of any water problem in the municipality, in public space, call your water company.*

This option to integrate the full urban water cycle and undertake more responsibilities to a higher standard without increasing tariffs also had a significant positive impact on the water company image. In the eyes of public opinion and the media, the water company became associated with attractive environmental projects, such as rehabilitated urban streams, leisure footpaths and beaches. Many thousands of people were given the possibility to walk along the urban streams and to swim in Blue Flag quality beaches for free. In addition, activities associated with the urban streams and beaches helped 100,000 customers to make the decision to voluntarily connect their properties to the new sewerage network over six years. A comprehensive environmental education program was undertaken to pass a simple and very effective message: "*connect to the sewerage system to have clean rivers and Blue Flag beaches*". These new activities, including the stormwater system transferred from the municipality to the water company, were financed by efficiency and productivity gains.

The organizational chart, typical of a hierarchical public service, with many intermediate levels of decision, has been simplified, with only one level of decision below the Board of Directors (Figure 5.2.1).

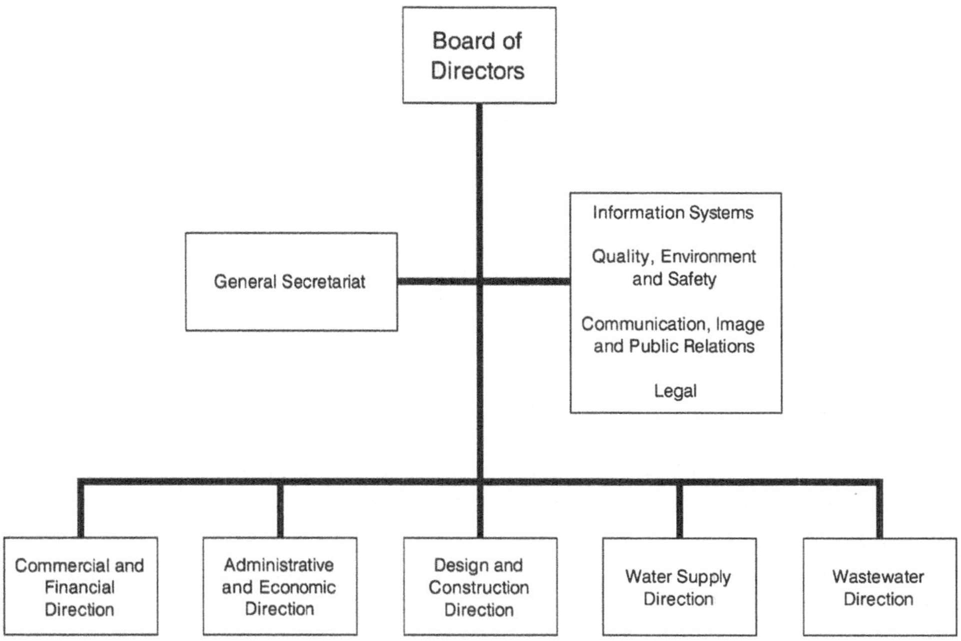


**Figure 5.2.1** New organizational chart.

Ensuring the stability and reassuring the workforce in the initial stages of the process of change was considered a priority, as it was vital to maintain the water and wastewater systems running as usual, without any negative impact on customers. Dialogue with the workers created a climate of trust and minimized the inevitable anxiety associated with change.

Over the first months, an attempt was made to identify the best workers; some of them were invited to assume more demanding, coordinating or managerial responsibilities and offered pay rises. During the first year, around 10% of workers were given a promotion based on merit. An annual bonus, worth on average an extra month of salary (varying individually from zero to three months salary), was awarded during the first year of operation, as a share of the profits due to higher productivity.

Customer relationships were given special attention and workers were invited to adopt a new, friendlier and more cooperative attitude towards them. Another interesting aspect which contributed to increasing customer satisfaction was the ability to solve almost all technical problems, such as bursts, lack of pressure or leaking meters on the same work day on which they were reported.

## 5.2.3 NON-REVENUE WATER REDUCTION

Non-Revenue Water (NRW) was initially reported as 28,085 $m^3$/day, equating to around 47% of the System Input Volume. The inherent knowledge of the system was not sufficient to produce a reliable water balance or to quantify the division between real and apparent losses.

However, there was clearly a high proportion of real losses given the many visible leaks and the high number of bursts. Therefore the first priority was a faster repair of visible leaks and bursts, not only to reduce NRW but also to improve the public image of the water company: people were accustomed to seeing valuable water flowing in the streets for days, weeks and even months and nobody bothered to inform the water company about it.

Active leakage control teams were created at an early stage to locate invisible leaks and it was surprising to notice the capability of a high number of workers to effectively use listening sticks.

There was little scope, initially, for pressure reduction because of local topography and because many buildings had up to ten floors and did not have their own pumping systems.

Later, the route and branch design of the network was inter-connected to a higher level with greater redundancy and it was possible to optimize customer connections to the most appropriate water reservoirs or supply zones, reducing the pressure in many pipes. Simultaneously, a more active approach was taken towards customers who were failing to pay their bills or who had illegal connections and faulty meters.

The volume of NRW was almost halved from 28,085 to 14,305 $m^3$/d during the first two years of the change process without any mains replacement program or widespread meter exchange project. Between 1999 and 2000, the water balance highlighted marked differences resulting from the first actions to reduce NRW, as shown in Figure 5.2.2. Between 1998 and 2004 there was a substantial reduction of NRW as shown in Figure 5.2.3.

It is interesting to note that between 1999 and 2004 the water company replaced only 5% of the mains in the distribution system. The mains selected for replacement were mainly old asbestos cement pipes in the more densely populated zones, mainly to reduce bursts. The remaining network was installed recently, with the most common material being PVC in the distribution mains and ductile iron in the larger diameter mains. In order to reduce apparent losses, a program was developed to replace all the meters over ten years old.

These measures were sufficient to reduce NRW from 28,085 $m^3$/day to 11,024 $m^3$/day in six years (a reduction of 61%), with annual savings equivalent to 6% of the average annual operational turnover.

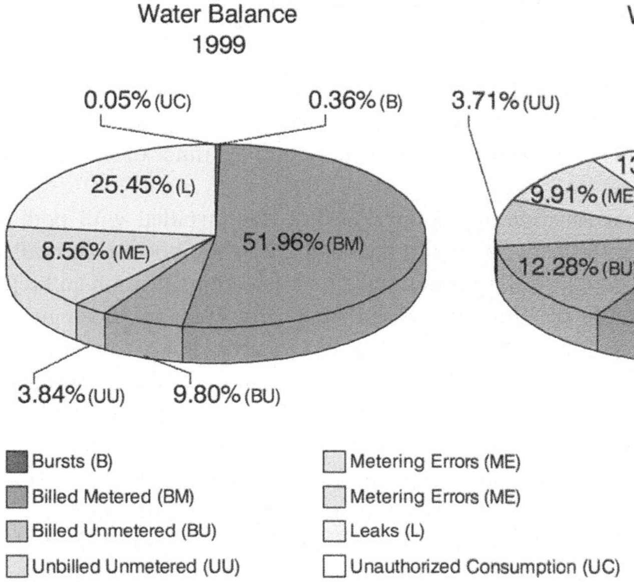


**Figures 5.2.2** Water balance in 1999 and 2000.

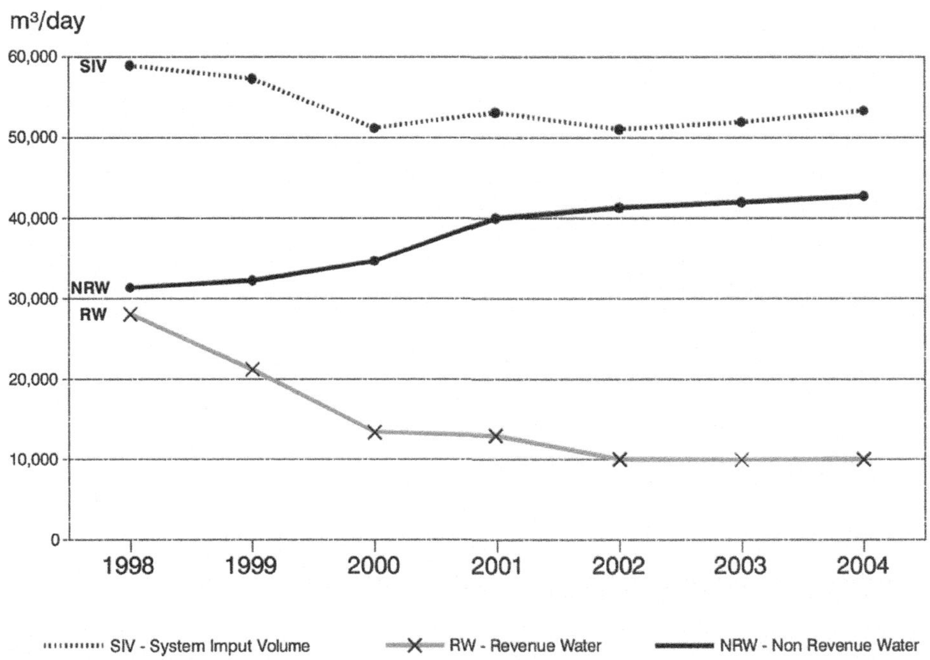


**Figure 5.2.3** Evolution of the System Input Volume, Revenue Water and Non-Revenue Water of *Águas de Gaia* (1998-2004).

## 5.2.4 MONITORING THE WATER SUPPLY SYSTEM

In 1997, water supply system monitoring was rudimentary, with, for instance, no meters in the water reservoirs, no telemetry or control room available 24 hours a day. One of the first priorities was to install flow meters at the water reservoirs to make it possible to estimate the night flows and detect bursts at an early stage. In the beginning the meters had to be read by an operator at fixed times of the day, whilst later, a telemetry system was installed.

At the same time, a Control Room functioning 24 hours a day was installed with both telephone operators and technically trained staff. In the beginning, it had a simple white board on a wall on which bursts and other anomalies were recorded, including communication, detection, location and repair times. There was no formal telemetry system at the time: the service reservoirs were visited regularly one or two times a day. Flows from the reservoirs and their respective levels were later automatically received, by telemetry in the Control Room and systematically analysed and displayed. Simple comparison with historical data collected at the same hour and day of the week made it possible to detect anomalies of consumption, namely bursts.

During normal working hours, information relevant to network operations was centralized in the Control Room and immediately transmitted to the operating structure for swift identification of the causes of anomalies and their fast and effective repair. Thus, reliability of the system showed a marked improvement, as revealed by the constant decline in supply interruptions and the increase of the percentage of interruptions resolved in less than an hour (Figure 5.2.4).

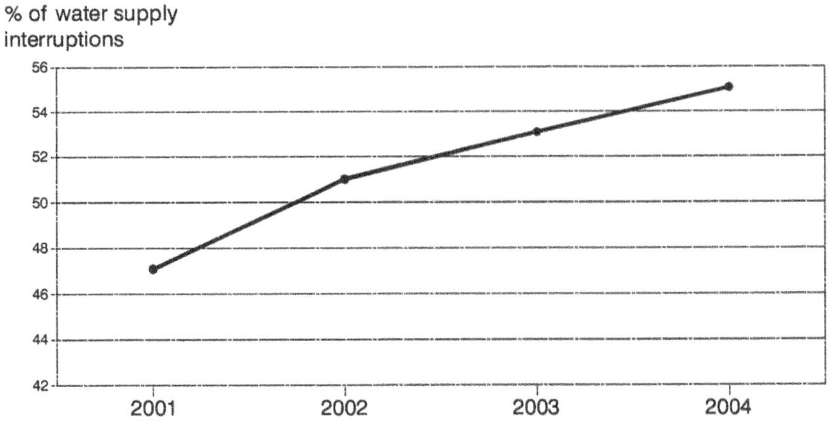


**Figure 5.2.4** Water supply interruptions resolved in less than one hour.

## 5.2.5 BUILDING A NEW WASTEWATER SYSTEM

In 1997 there were no wastewater treatment plants in operation in the municipality and the sewerage system covered less than 10% of the area. The population used septic tanks, many of them connected to the stormwater system or to urban streams, risking both public health and the environment.

In just six years, the company was able to fulfil its main goal in this area, endowing the municipality with a modern and efficient wastewater system (Figure 5.2.5).

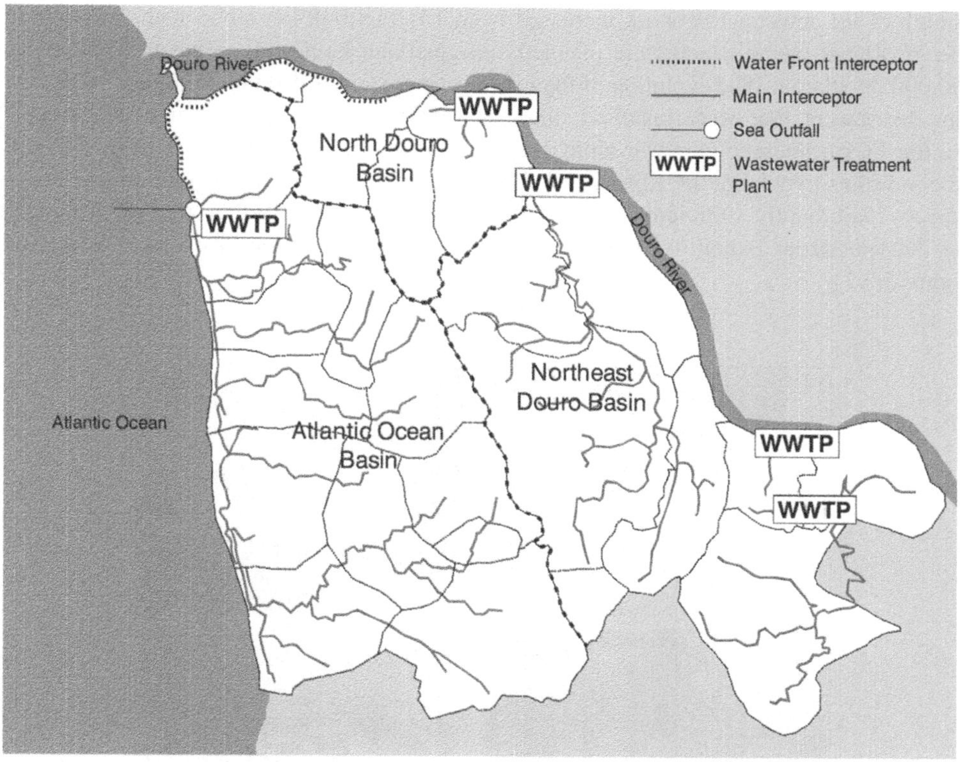


**Figure 5.2.5** The new wastewater system.

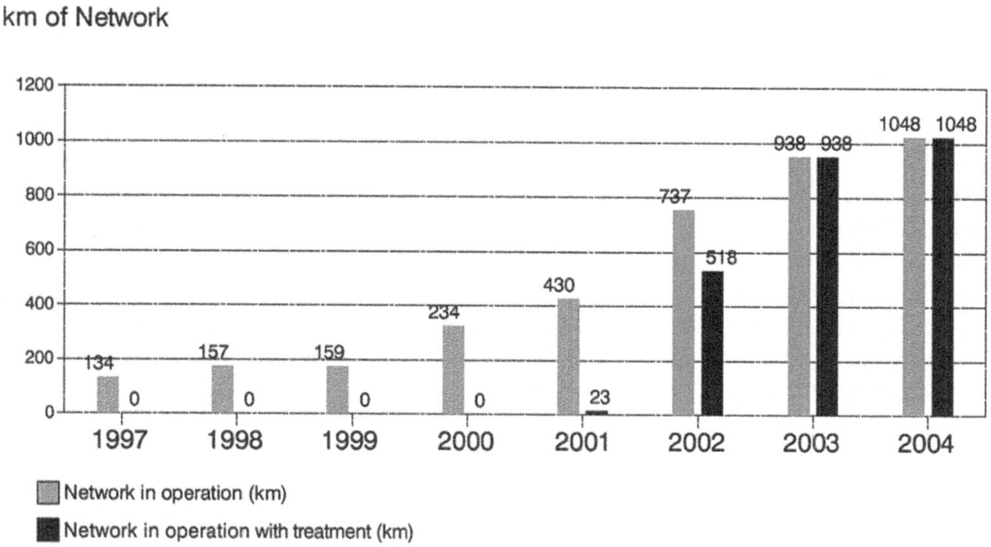


**Figure 5.2.6** Evolution of the wastewater system (1997-2004).

The length of the sewerage network increased from 157 to 1,048 km along with five state of the art wastewater treatment plants which were planned, designed and built with sufficient capacity to serve a future 500,000 population. The evolution of the wastewater system is detailed in Figure 5.2.6.

Among the sewers that were installed, there is a less conventional one that deserves a special reference: the 11 km interceptor sewer along the left bank of the River Douro, built in such a way that it created conditions to enlarge the adjacent road and to create a footpath, where there was formerly just a very narrow road, barely sufficient for a small car to pass safely. Installation of this interceptor was critical to the waterfront requalification, transforming it into one of the city's most attractive leisure areas (Figure 5.2.7).

**Figure 5.2.7** Gaia waterfront interceptor.

No matter how important and difficult, simply installing the sewers was only part of the solution as it is also essential that properties are actually connected to the system. This was not easy because many people had only recently invested in building septic tanks and thought that their problems were solved. Therefore, a very active marketing operation was launched to inform the population about the need to connect existing buildings to the new sewers. The main message was: "*connect your house to the sewer to have clean rivers and Blue Flag beaches*". The population responded very well to the request, as shown in Table 5.2.2. The launch of a special credit line (Figure 5.2.8) to finance necessary works that were directly and indirectly related to connections to the sewerage network was essential in facilitating the connection of up to 3,000 customers (2.5% of the total potential customers in the municipality) per month.

The activities and responsibilities of *Águas de Gaia* were extended in 2002 to include operation of the stormwater network. This new activity began with preventive maintenance of the network during the summer months, using workers and equipment from the wastewater department of the company. This resulted in almost no additional costs. It was, however, known that management of the stormwater system was likely to result in significant future investment costs that were difficult to anticipate. In fact, the stormwater system was much older than the water supply and wastewater systems and there was insufficient

**Table 5.2.2** New connections to the sewerage network (1998-2004).

| Year | Nº of Customers | Growth | |
|---|---|---|---|
| | | New connections | % |
| 1998 | 18,671 | - | - |
| 1999 | 29,934 | 11,263 | 60.3 |
| 2000 | 38,700 | 8,776 | 29.3 |
| 2001 | 52,377 | 13,677 | 35.3 |
| 2002 | 70,887 | 18,510 | 35.3 |
| 2003 | 94,495 | 23,608 | 33.3 |
| 2004 | 105,420 | 10,925 | 11.6 |

**Figure 5.2.8** Leaflet for a credit line for home connection to the sewerage network.

and inaccurate information in the infrastructure databases in terms of location, materials, age and condition of the assets.

At the time, there was no municipal tax to cover the stormwater service and it was not a political option to create it. Therefore, any new investment in maintenance and expansion of the system had to be obtained through efficiency gains of the water company and paid by customers through the water and wastewater tariffs.

## 5.2.6 HIGH INVESTMENT EFFORT

Until 2004, the total investment by *Águas de Gaia* amounted to 150 million euros, of which more than 123 million Euros was directed towards the sewerage system. About 60% of the investment was financed by funds generated by the company, namely from savings resulting from water loss reduction and tariffs. The remaining 40% came from European funds. Figure 5.2.9 outlines the evolution of investments over the years, 85% of which refer to the wastewater system.

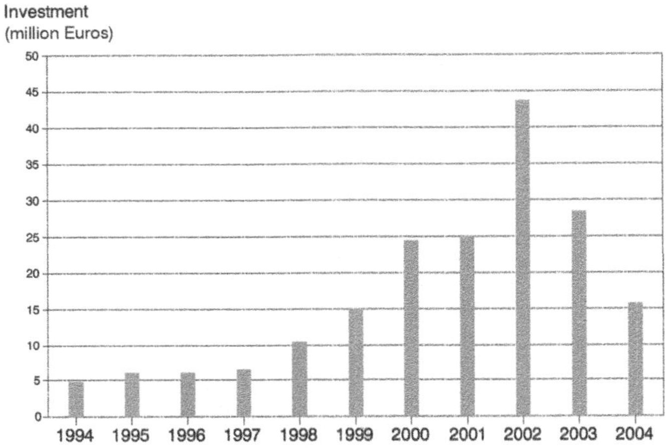


**Figure 5.2.9** Evolution of investment in period 1994-2004.

## 5.2.7 CLEANING, OPENING AND REHABILITATING THE URBAN STREAMS

The excellent performance of the wastewater system was reflected in the improvement of water quality in urban streams, rivers and beaches (Figure 5.2.10). In 2001, the municipal company initiated rehabilitation of urban streams, involving three phases: eliminating pollution, opening streams if covered and rehabilitating banks. In Gaia, there were about 400 km of urban streams, the majority of which were grossly polluted and showed severe signs of degradation.

**Figure 5.2.10** Example of a stream in Gaia before and after rehabilitation.

The strategy to remove pollution from these streams included the installation of sewers in the streets whenever possible, but in many cases it was necessary to install them in the banks of the urban streams to intercept existing discharges of sewage. The fact that the main outfalls were located on the banks of the streams was a main cause of pollutant discharges. The stream degradation process is illustrated in Figures 5.2.11 to 5.2.16.

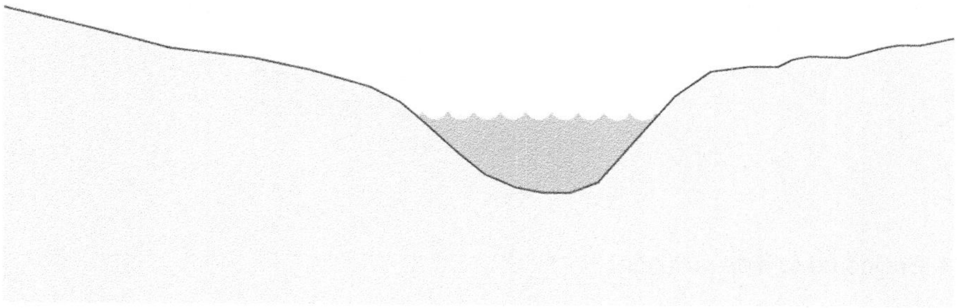

**Figure 5.2.11** Watercourse in its natural state.

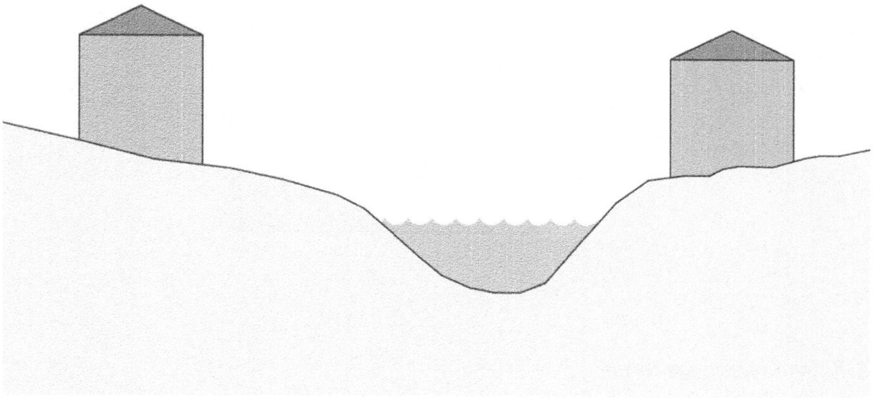

**Figure 5.2.12** Improper construction on the banks.

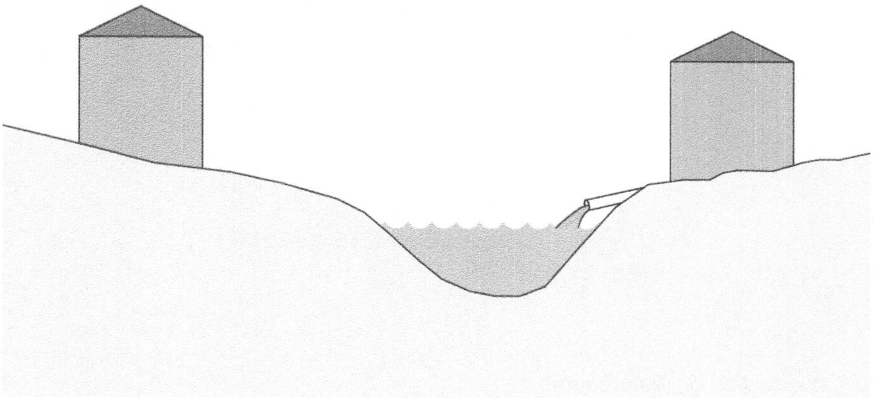

**Figure 5.2.13** Pollution caused by sewage discharge.

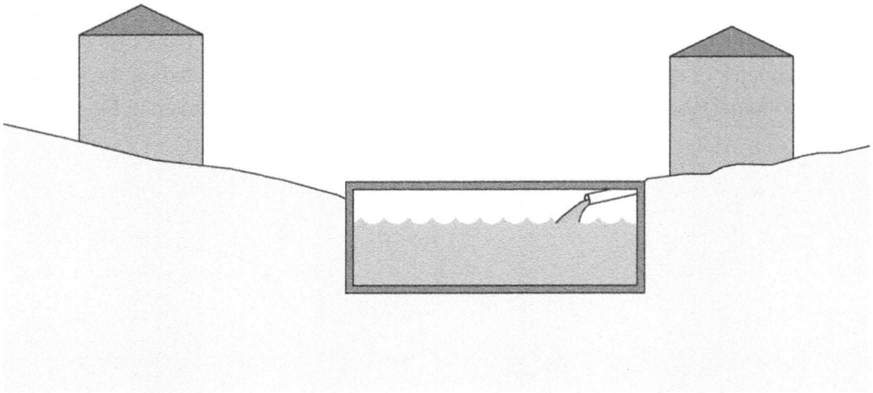

**Figure 5.2.14** Enclosure to hide pollution.

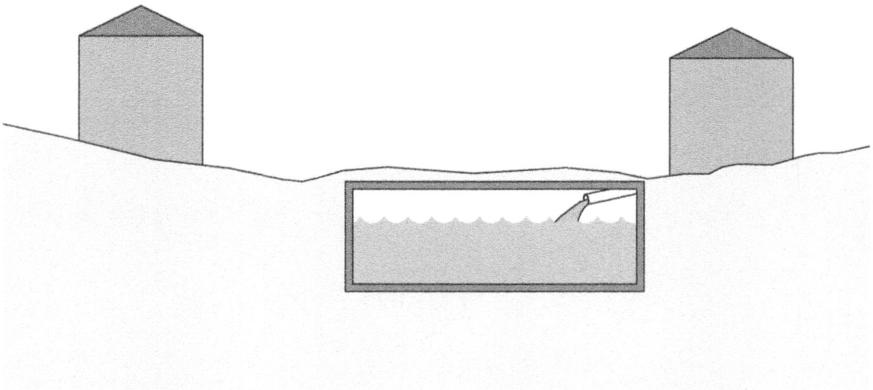

**Figure 5.2.15** Total coverage of water.

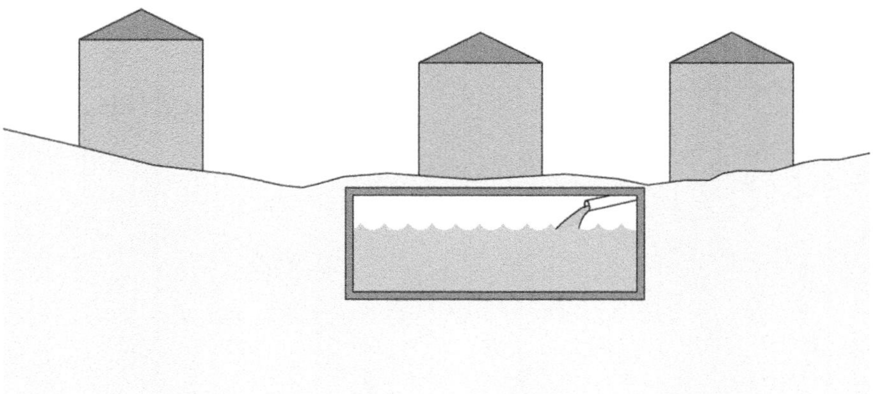

**Figure 5.2.16** Construction on watercourse.

However, it was essential to initiate any intervention by cleaning the bed and banks of the watercourses in order to remove the huge amounts of rubbish and weed vegetation. In many situations, only after the first, necessarily simple, cleaning process was it possible to assess the stream and to set the type and extent of the complementary intervention required (Figure 5.2.17). Further cleaning and maintenance operations were required until the population was finally persuaded to adopt positive attitudes towards defence of the stream, developing an emotionally protective attitude that was crucial to avoid vandalism.

The construction of walkways on the banks was crucial for mobilizing support by the local population for the project, whilst this mobilization also helped to overcome the natural resistance of some property owners to authorize the use of their land to install public footpaths. These paths allowed faster, secure, car-free access to the beaches and combined with the 17 km walkway that runs parallel to the coastline.

In addition to constant monitoring of watercourses to prevent anomalies and identify possible polluters, the company cleaned and maintained approximately 130 km of streams and watercourses until 2004, developed requalification projects in 10 km of streams, including walkways and leisure spaces along the banks. Interestingly, many doctors, mainly cardiologists, were very important for the project, "forcing" their patients to walk on the healthy footpaths, with the message "move for your health" passed to all the population in pamphlets distributed with water bills.

**Figure 5.2.17** Cleaned and rehabilitated urban streams.

## 5.2.8 BLUE FLAG BEACHES

Water quality improvements were also the starting point for requalification of the coastline and seafront. The coast of Gaia became one of the most dynamic and attractive centres of urban leisure and recreation in the greater Porto region. In 2002 and as a result of satisfactory analytical results of bathing water quality, Gaia was eligible to apply for the Blue Flag award. In that year, the Blue Flag was hoisted in twelve bathing areas, a process that culminated in the attainment of this distinction for all beaches in 2004 (Figure 5.4.18).

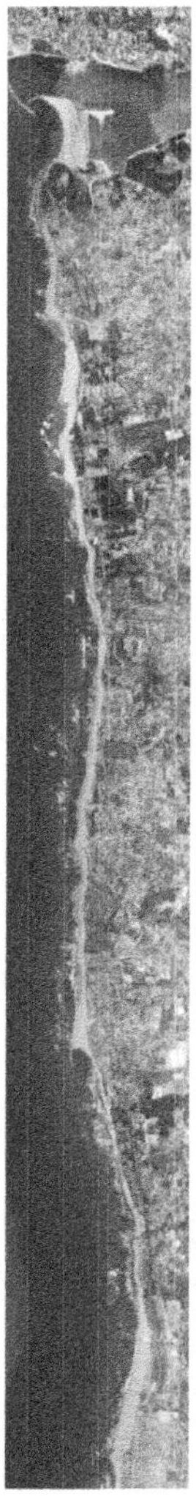

| | 1993 | 1994 | 1995 | 1996 | 1997 | 1998 | 1999 | 2000 | 2001 | 2002 | 2003 | 2004 |
|---|---|---|---|---|---|---|---|---|---|---|---|---|
| 01 | ✓ | ✓ | × | ✓ | × | × | ✓ | ✓ | ✓ | ✓ | Blue Flag | Blue Flag |
| 02 | × | ✓ | × | ✓ | × | × | ✓ | ✓ | ✓ | ✓ | Blue Flag | Blue Flag |
| 03 | - | - | - | - | - | - | - | - | - | Blue Flag | Blue Flag | Blue Flag |
| 04 | - | - | - | - | - | - | - | - | - | Blue Flag | Blue Flag | Blue Flag |
| 06 | × | × | × | ✓ | × | × | ✓ | ✓ | ✓ | ✓ | Blue Flag | Blue Flag |
| 07 | - | - | - | - | - | - | - | - | - | Blue Flag | Blue Flag | Blue Flag |
| 08 | - | - | - | - | - | - | - | - | - | Blue Flag | Blue Flag | Blue Flag |
| 09 | × | ✓ | × | ✓ | × | × | ✓ | ✓ | ✓ | Blue Flag | ✓ | Blue Flag |
| 10 | - | - | - | - | - | - | - | - | - | Blue Flag | Blue Flag | Blue Flag |
| 11 | ✓ | ✓ | ✓ | ✓ | × | ✓ | ✓ | × | ✓ | Blue Flag | Blue Flag | Blue Flag |
| 12 | - | - | - | - | - | - | - | - | - | ✓ | Blue Flag | Blue Flag |
| 13 | - | - | - | - | - | - | - | - | - | Blue Flag | Blue Flag | Blue Flag |
| 14 | - | - | - | - | - | - | - | - | - | Blue Flag | Blue Flag | Blue Flag |
| 15 | × | × | × | ✓ | × | × | × | × | ✓ | Blue Flag | Blue Flag | Blue Flag |
| 16 | - | - | - | - | - | - | - | - | - | Blue Flag | Blue Flag | Blue Flag |
| 17 | ✓ | × | × | × | × | × | × | Blue Flag | ✓ | Blue Flag | Blue Flag | Blue Flag |
| 18 | × | ✓ | × | ✓ | × | × | ✓ | × | ✓ | ✓ | Blue Flag | Blue Flag |

- Unclassified area (without information)
× Not meeting the Bathing Water Standards
✓ Meeting the Bathing Water Standards
Blue Flag

**Figure 5.2.18** Evolution of the classification of the urban beaches.

Gaia has become the Portuguese municipality with the greatest number of Blue Flag beaches. This process led to a series of significant improvements along the seafront, especially construction of facilities for sport and leisure, including a pedestrian walkway that connected all the beaches (Figures 5.4.19 to 5.4.21). The effect of this policy was to create a significant attraction towards the sea that went beyond the municipality boundaries and reached people from neighbouring areas, creating a rediscovered and important centre for leisure and free-time.

**Figure 5.2.19** Beaches of Gaia – Blue Flag.

**Figure 5.2.20** Pedestrian walkway along the coastline and a sports park.

**Figure 5.2.21** Pedestrian walkway and cycleway along the coastline.

## 5.2.9 ENVIRONMENTAL EDUCATION

A few minutes from one of the best beaches, the Gaia Urban Streams Environmental Education Centre was built in 2002 (Figure 5.2.22). The centre aimed to develop educational activities and disseminate the environmental strategy of the municipality towards both children and adults. The aim was to raise awareness amongst visitors about their environmental responsibilities, particularly concerning issues related to water, waste and nature conservation. The environmental education program, originally designed for local schools, has attracted visitors from schools from all over the country.

It is very important to realize that this environmental education centre was initially regarded mainly as an investment to attract more customers. As previously mentioned, a simple message was passed to teachers and school kids and through them, to parents and grandparents: namely, *connect to the sewage system to benefit from clean streams and Blue Flag beaches*. In fact, over 100,000 homes, 80% of the total in the municipality have been connected to the new sewerage system in just six years.

Seeing many thousands of people moving along the walkways and using the beaches was certainly rewarding, as was the result of a survey of the population of the municipality in 2003: the people had to choose a symbol to represent the municipality with three possible choices: Porto Wine (which is, indeed, matured in Gaia), the River Douro and the beaches. Surprisingly, although the Porto Wine and the Douro were there for many, many years, and the Blue Flag beaches only for three years, the population chose as a municipal symbol: "Gaia, city of beaches".

**Figure 5.2.22** Environmental education centre dedicated to the rehabilitation of urban streams.

## 5.2.10 REFERENCES

*Águas de Gaia* (1999). Annual Report 1998 (Portuguese Version). *Águas de Gaia*, Vila Nova de Gaia, Portugal.
*Águas de Gaia* (2000). Annual Report 1999 (Portuguese Version). *Águas de Gaia*, Vila Nova de Gaia, Portugal.
*Águas de Gaia* (2001). Annual Report 2000 (Portuguese Version). *Águas de Gaia*, Vila Nova de Gaia, Portugal.
*Águas de Gaia* (2002). Annual Report 2001 (Portuguese Version). *Águas de Gaia*, Vila Nova de Gaia, Portugal.
*Águas de Gaia* (2003). Annual Report 2002 (Portuguese Version). *Águas de Gaia*, Vila Nova de Gaia, Portugal.
*Águas de Gaia* (2004). Annual Report 2003 (Portuguese Version). *Águas de Gaia*, Vila Nova de Gaia, Portugal.
*Águas de Gaia* (2005). Annual Report 2004 (Portuguese Version). *Águas de Gaia*, Vila Nova de Gaia, Portugal.

# Chapter 5.3

# Case study 3 - *Copasa* (2003-2013)

*With Ricardo Augusto Simões Campos*
*CEO of* Copasa *(since 2009)*

## 5.3.1 INTRODUCTION

*Copasa* is one of the largest water companies in Brazil and, indeed, in the world. In terms of ownership, it is 51% public, publicly managed and listed on the São Paulo Stock Exchange (BM&F BOVESPA). In 2003, the company and the whole state of *Minas Gerais* were subject to a change management process known locally as Shock Management. This process had the very strong political commitment of the State Governor himself.

Very impressive results were achieved from the first year onwards, whilst the process included widespread participation of company employees at all levels, high visibility with the local population and media, and a significant social component, namely through specific attention to low income customers. The change management process has already reached a mature status, with consolidation of results and procedures.

*Copasa* is a very interesting example of a company that is publicly controlled and managed, but with the distinctive advantage of having private shareholders, in terms of financing, transparency and impact mitigation of political cycles on operational management. Some features of this large Brazilian water company may be useful to others all over the world: the model of public ownership and public/private governance, the sustainable profitability, the social concern with access to water services, the strict control of non-payment of bills, the variable component of employee salaries and the culture of management excellence at all levels of the company, acknowledged through national and international prizes.

The state of *Minas Gerais* has the country's second largest population with 20 million people, equivalent to 10 percent of Brazil's population. Covering almost 600,000 km$^2$, it is the fourth most extensive state in the country and its economy is the third largest in Brazil (Figure 5.3.1). Water supply and sanitation services are managed by *Copasa*, a publicly-held mixed capital company, listed on the New Market, the top segment of corporate governance of BM&F BOVESPA (Securities, Commodities & Futures Stock Exchange), with the State of *Minas Gerais* as the controlling shareholder.

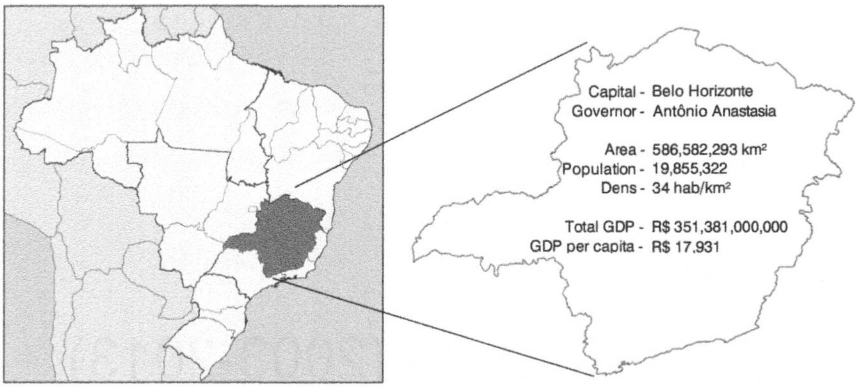


**Figure 5.3.1** State of *Minas Gerais* in Brazil.

In 2012 the company supplied water to a population of more than 14 million people, which represented 73% of the State of *Minas Gerais,* total population, while the population coverage by the sewerage network was approximately 9 million people. In 2003, *Copasa* engaged in the new Shock Management project of the state of *Minas Gerais*, with the aim to "become the best water company in Brazil" (Figure 5.3.2). Since 2003, the company has undergone a broad and deep process of change, in order to achieve high levels of operational, commercial and financial performance. Good results were observed from the first year of this process: the company went from deficit to profit in less than a year and it has been subsequently recognized as one of the best public utility services companies in Brazil.

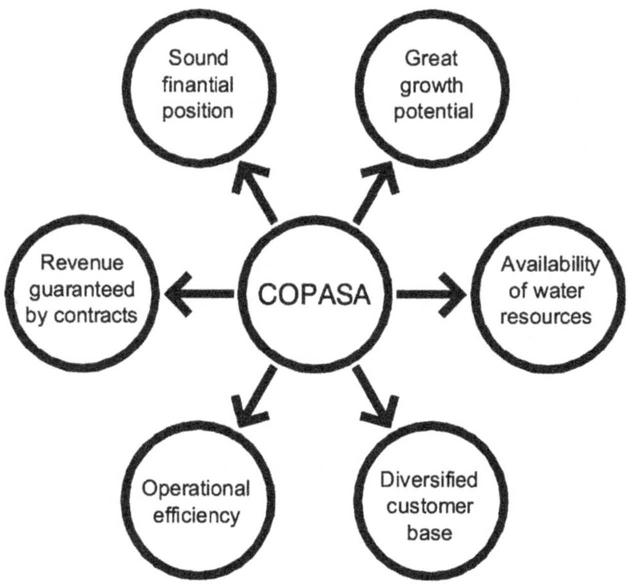


**Figure 5.3.2** Main strengths of *Copasa*.

## 5.3.2 FIFTY YEARS OF HISTORY

Until 1963, *Minas Gerais* had no formal sanitation policy, with the condition of both water supply and sanitation services being far from ideal. In 1963, the State Government created *Comag – Companhia Mineira de Água e Esgotos* (the Water and Sewage Company of *Minas Gerais*) – to define and implement a policy of basic sanitation covering water supply and waste water for the State of *Minas Gerais* (Figure 5.3.3).

| | |
|---|---|
| 1963 | The State Government creates Companhia Mineira de Água e Esgotos, COMAG. |
| 1966 | The basic sanitation services, which were in charge of the City Hall of Belo Horizonte, started being under the responsibility of the Municipal Department of Waters and SEWAGE, which had just been created. |
| 1970 | Incorporation of the activities of DEMAE to COMAG. |
| 1971 | DEMAE joins COMAG taking advantage of the federal ressources to be transfered by means of the creation of the National Sanitation Plan, PLANASA. |
| 1973 | COMAG starts being the promoting body of PLANASA in all the State of Minas Gerais.<br>COMAG enters into a concession agreement of water and sewage services of Belo Horizonte. |
| 1974 | COMAG transformed into Companhia de Saneamento de Minas Gerais (COPASA). The company serves 1.2 million people with treated water in 39 locations. The wastewater services benefit 804 thousand people in 5 locations. |
| 1982 | 5 million people, in all the State of Minas Gerais, are supplied with water treated by COPASA. |
| 1990 | COPASA creates the Wellsprings Protection Integrated System - SIPAM, an environmental education program to combat the degradation of river basins. |
| 1994 | COPASA celebrates 20 years and is present in 426 municipalities in Minas Gerais, supplying with treated water a population of 7.4 million people. In terms of wastewater it serves 3.1 million peolple. |
| 2001 | 10 million people receive treated water from COPASA |
| 2003 | COPASA reaches the first positive financial result of R$94.1 million |
| 2005 | The company files with the Brazilian Securities and Exchange Comission a registration request for the primary public offering of shares. |

**Figure 5.3.3** Fifty years of history of *Copasa*.

In 1971, another important step was taken, when the Federal Government of Brazil created the National Sanitation Plan (*Planasa*), which set targets to be achieved by the country regarding basic sanitation. During that same period, the Municipal Water and Sewage Department of *Belo Horizonte*, responsible for providing these services in the city, was incorporated into *Comag*, taking advantage of federal funds to be transferred through the *Planasa*. Comag was transformed into *Copasa MG –Companhia de Saneamento de Minas Gerais* [Sanitation Company of *Minas Gerais*].

*Copasa*'s main activity is the provision of water supply and sewage services to the population, which includes the planning, design, construction and management of projects along with expansion and operation of the systems. The company accumulated losses for eleven years, until the State Government of *Minas Gerais* decided to adopt a shock management process. Since 2003, *Copasa* has recorded a history of positive operational and financial results. In 2006, *Copasa* undertook its Initial Public Offering (IPO) entering into the BM&F BOVESPA New Market. In 2008, a secondary offer of shares was made, in which the *Belo Horizonte* Municipality disposed of all of its shares and the State of *Minas Gerais* sold part of its shares without losing corporate control of the company.

## 5.3.3 COMPANY PROFILE

In 2002, *Copasa* had concession contracts to provide water supply in 593 municipalities, serving 10.4 million customers, in addition to concession contracts to provide sewage services in 138 cities, totalling 4.8 million customers. In 2013, the company operated in 625 of the 853 municipalities of *Minas Gerais*: 346 cities with water supply concessions and 279 cities with water supply and wastewater concessions. By 2025, the aim is to reach all municipalities in the State of *Minas Gerais*.

In the municipalities where it operates, *Copasa* serves 98% of the population with water services and 83% with wastewater services (Figure 5.3.4). The evolution of *Copasa*'s performance indicators are presented in Figure 5.3.5 (*Copasa*, 2009 and 2013).

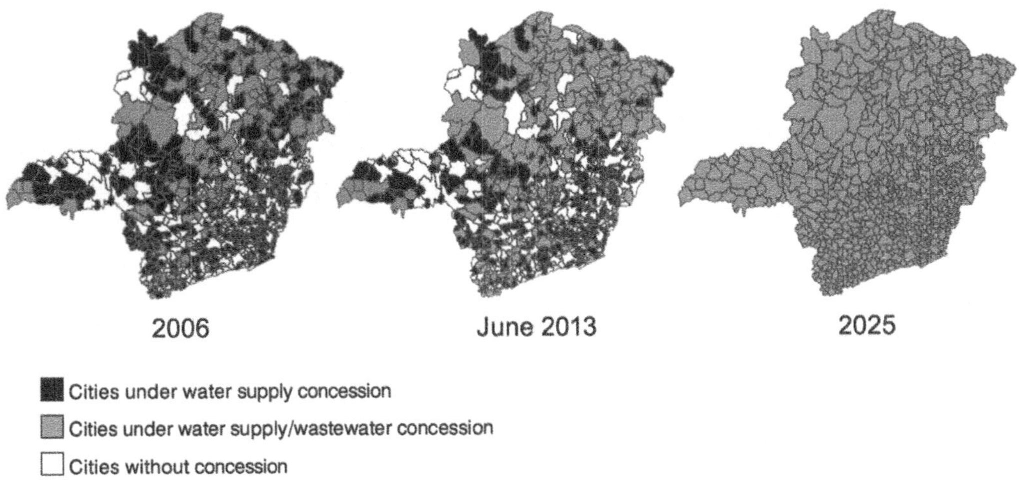


**Figure 5.3.4** Concessions for water supply and sewage in the State of *Minas Gerais*.

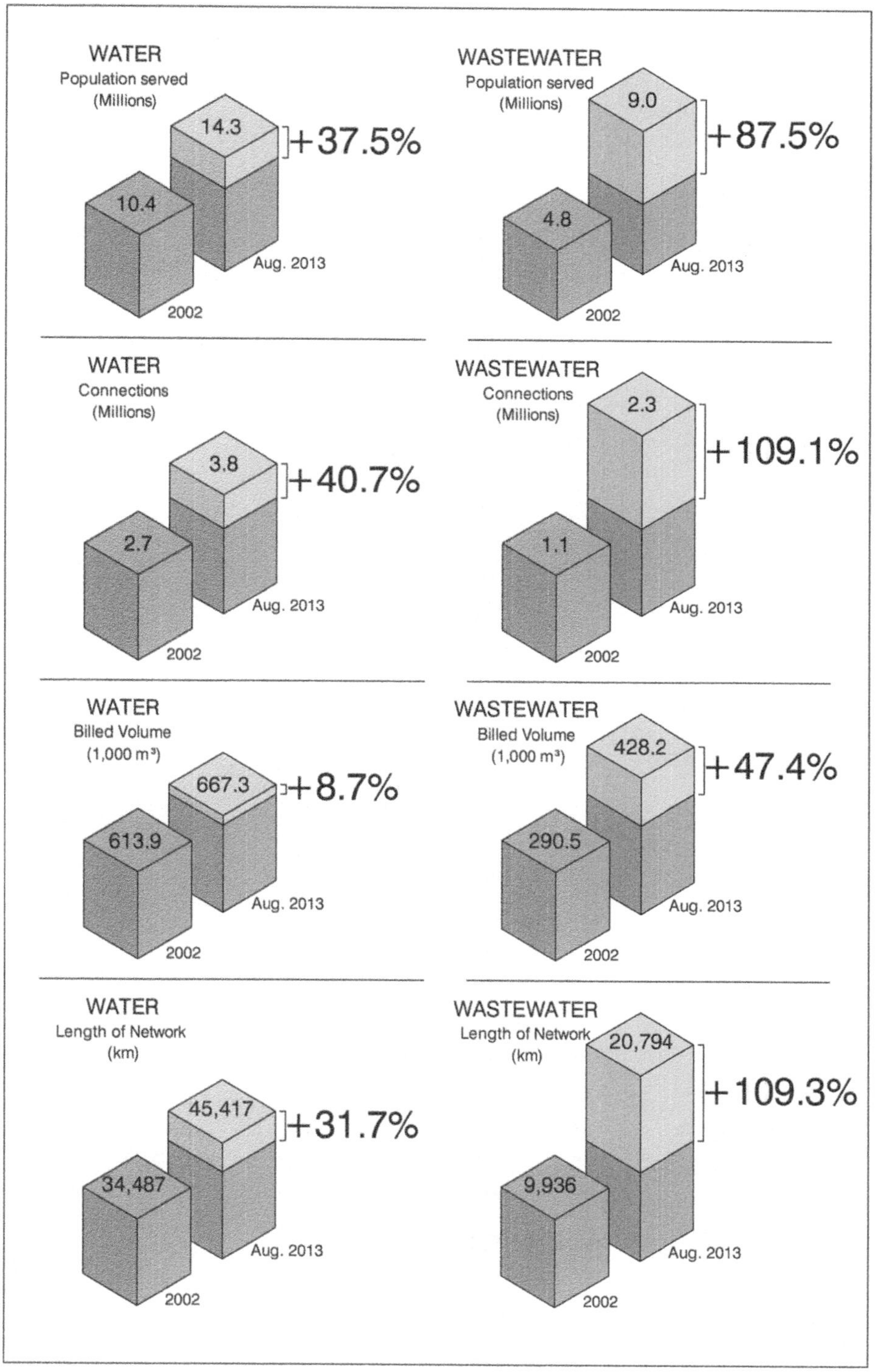


**Figure 5.3.5** Evolution of water provision and sewage coverage (Source: *Copasa*, 2013).

Between 2003 and 2012, the company invested R$ 6.8 billion (as of March 2014, R$ 1 = € 0.31), especially for the expansion and improvement of water supply and waste water systems (Figure 5.3.6). The evolution of annual investments is presented in Figure 5.3.7.

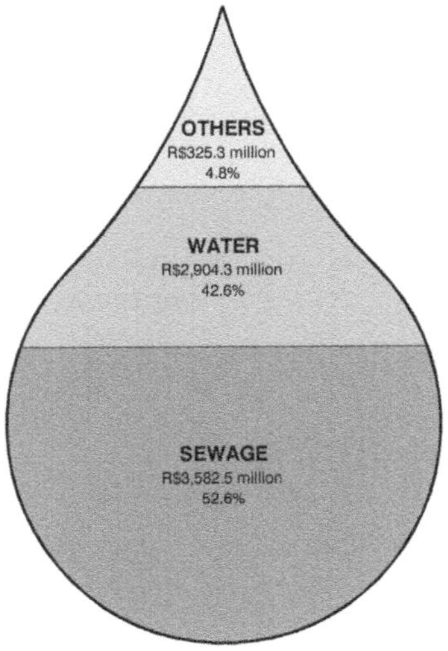


**Figure 5.3.6** Distribution of total investment (2003-2012).

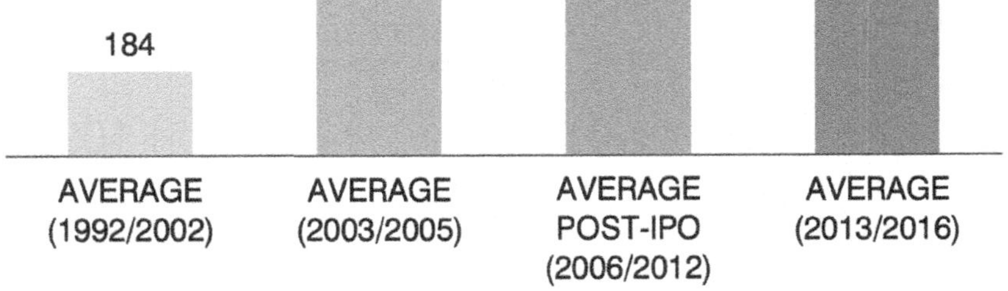


**Figure 5.3.7** Evolution of annual investment (1992-2016).

## 5.3.4 SHOCK MANAGEMENT IN *MINAS GERAIS*

In 2003, the major political concerns were the fiscal crisis; the State's public sector institutions and procedures were obsolete and inefficient, and the government started implementation of Shock

Management. In 2007, the Government shifted its focus towards improving the quality of public services and, from 2011, has been increasingly concerned with ensuring that achievements are sustainable over time and that citizens have an adequate voice in the State's policy decisions. These three generations of reforms, large scale examples of change management processes, are identified respectively by the following names: "Shock Management", "State for Results" and "Management for Citizenship" (Anastasia, 2009).

### 5.3.4.1 Shock Management (2003-2006)

Upon taking office in 2003, the new Government of *Minas Gerais* faced a severe fiscal crisis, but also recognized that the State's public sector institutions were dysfunctional. Reorganizing the state and resolving the fiscal crisis were considered to be pre-conditions to restoring the state's ability to provide quality public services and to foster economic development.

Shock Management established a new organisational model and accountability mechanisms, whose keywords were "people, quality and innovation". This true shock therapy aimed to bring public spending under control and lay institutional foundations for better public sector management. Traditional fiscal adjustment measures were combined with innovative activities to modernize the underlying administrative structure and introduce a results-based management system (Figure 5.3.8).

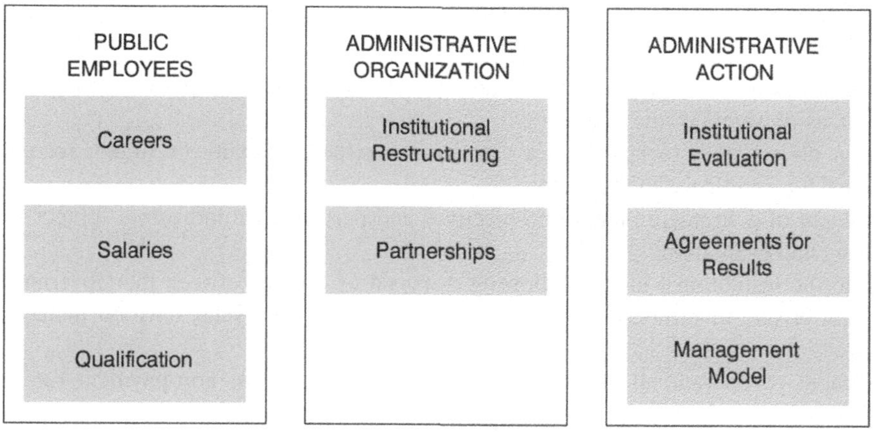


**Figure 5.3.8** The Shock Management model.

Several unpopular measures were taken to cut expenses, including the elimination of 65 public entities, cutting 3,000 political jobs, payroll centralization, implementation of a new purchasing policy and debt renegotiation.

### 5.3.4.2 The State for Results (2007-2010)

By 2007, the Government had largely restored order to public finances, creating the basis for a second phase, referred to as the State for Results, aiming to improve service delivery and promote economic growth. A key Government objective was to apply management reforms in critical sectors to significantly improve public sector performance in the short term.

The State for Results reform was converted into the *Minas Gerais* Integrated Development Plan (PMDI), which extends until 2023. The PMDI has two pillars: a quality tax, namely spending better the public resources, as well as quality and innovation in public management.

#### 5.3.4.3 Management for Citizenship (2011-2014)

The Government's third phase of reform adds new dimensions and participatory arrangements to its management model. However, it continues to focus on the development of strategic management capacity and continued improvements in its core management functions for increased efficiency and effectiveness. In this phase, efforts towards the achievement of results have a greater focus on the demand side, with delegation of key roles to lower levels of the administration, closer to the public.

#### 5.3.4.4 Shock management in Copasa

Amongst the measures taken, in addition to renegotiations with suppliers and spending cuts, a variable remuneration system for the company's 11,000 employees was implemented based on quarterly targets. The members of each team began to receive a bonus proportional to the percentage of goals achieved in their unit. The results were quick to appear. In just one year, *Copasa* moved from a loss of R$ 90 million to a profit of R$ 94 million.

The Shock Management strategic elements developed for the state of *Minas Gerais* were fully applied in *Copasa*, namely:

- Professionalization of management, beginning with the choice of a public manager rather than a politician, as president of the company;
- Definition of a vision: to be the best sanitation services company in Brazil, recognized as the benchmark for excellence in the sector;
- Development of a strategic plan with objectives and performance indicators subject to continuous monitoring and evaluation;
- Change to the institutional model, following dispersal of capital between the Government of *Minas Gerais* and private investors, avoiding the impact that political cycles have on management of the company;
- Improvement of operational efficiency: reduction of water loss, non-payment rates and energy consumption;
- Increase revenue through improved billing and collection for services provided and also through new municipal concession contracts;
- Balancing company accounts, which went from deficit to profit in less than a year;
- Involvement and motivation of employees in the company's strategic plan, introducing a variable component of remuneration linked to targets and results.

Another key aspect of the new *Copasa* management model relates to listing on the Stock Exchange. In February 2006, the company conducted its Initial Public Offering (IPO) and adhered to the highest level of corporate governance of BM&F BOVESPA – the New Market. To be listed in this segment, a company has to adhere to a set of corporate rules, known as good corporate governance practices, which are stricter than those established by Brazilian law. In 2012, the share price registered an increase of 35%, whilst, in the same period, the Ibovespa index increased only 7.4%. In 2013 the shareholder structure of the company became: 51% of shares held by the State of *Minas Gerais* with remaining shares (49%) in free float. In December 31, 2012, *Copasa*'s market value reached R$ 5.24 billion, with about 2,500 shareholders in 40 countries (Figure 5.3.9).

The organizational structure of *Copasa* is presented in Figure 5.3.10.

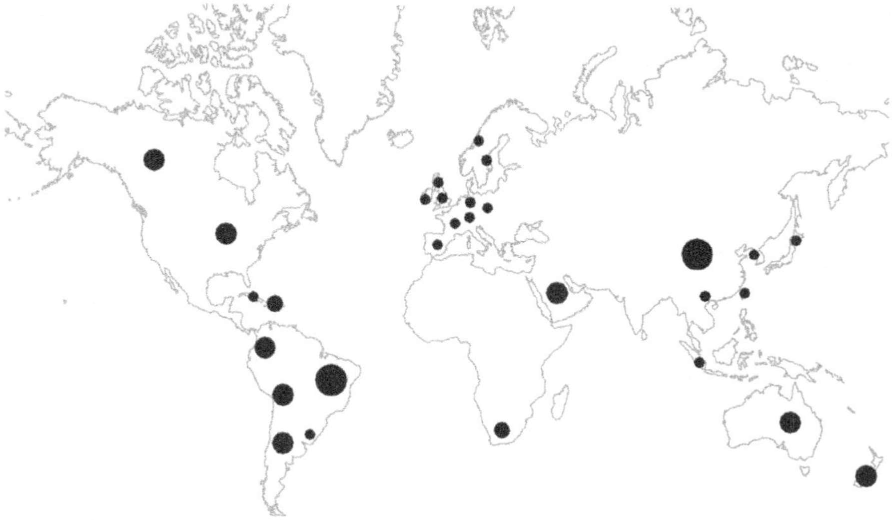

**Figure 5.3.9** Shareholders of *Copasa* in 40 countries.

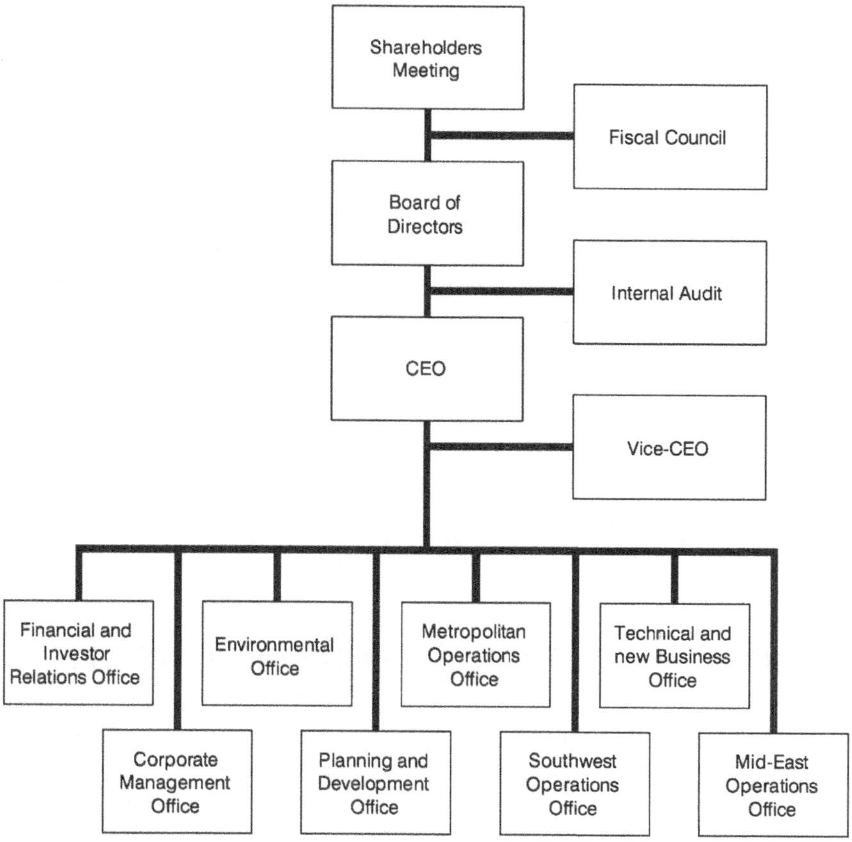


**Figure 5.3.10** Organizational structure of *Copasa*.

### 5.3.5 MANAGING FOR EXCELLENCE

*Copasa*'s Strategic Plan is the result of an extensive process of discussion and sharing of its vision for the future between employees, shareholders and customers, which established ten strategic objectives for the organization:

1. Permanent dialog with the granting authority;
2. Sustainable growth;
3. Socio-environmental responsibility;
4. Protection of water resources;
5. High level of Corporate Governance;
6. Regard for shareholder interests;
7. Valuation of employees;
8. Focus on customer satisfaction;
9. Quality of services provided;
10. Partnership with suppliers.

In 2010, the company was awarded the Gold Trophy and, in 2011, the Platinum Trophy, when it became the first Brazilian sanitation sector company to be recognized by ABES for the quality of its corporate management system.

Since 2005, the company has used the Balanced Scorecard (BSC) as a monitoring tool of the principal strategic challenges. The application of this methodology led to the creation of a Strategic Map for the company (Figure 5.3.11).

*Copasa* uses a variable remuneration model applied to all organizational units which rewards the collective endeavours of employees when defined performance indicators targets are reached. Accordingly, *Copasa* uses variable compensation models that reward the collective effort to meet the goals set in strategic planning:

- Institutional Performance Bonus (GDI): Quarterly evaluation system directly related to the strategic objectives of the company, rewarding employees according to the results obtained by their work unit each month. In 2012, the bonus percentage increased from 13.5% to 15.3% of employee basic salary.
- Performance Bonus for System Heads (GDES): Established in 2011, this bonus is granted to the employees working as System Heads and depends on the operating performance of sites under their responsibility.
- Management Performance Bonus (GDG): A bonus rewarding employees in management positions, depending on achievement of monthly pre-established goals;
- Profit Sharing (PL): Distribution of part of the company's annual net income to all employees, with each receiving an equal amount, regardless of pay level.

### 5.3.6 FINANCIAL PERFORMANCE: FROM DEFICIT TO PROFIT

*Copasa* moved from a R$ 90 million loss in 2012 to a R$ 94.1 million positive net financial result in 2013 and has been profitable ever since (Figure 5.3.12). The *Minas Gerais* company ended 2012 with net operating revenues for water and sewage services of R$ 2,800 million, an increase of 203% over the R$ 915 million reported in 2002 (Figure 5.3.13). The evolution of key economic financial indicators is presented in Table 5.3.1.

**Figure 5.3.11** Strategic Map for *Copasa*.

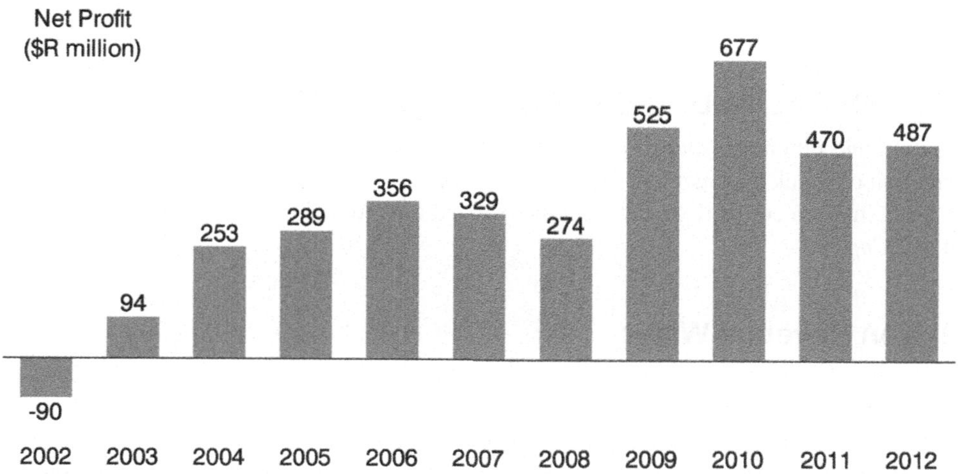


**Figure 5.3.12** Evolution of net profit.

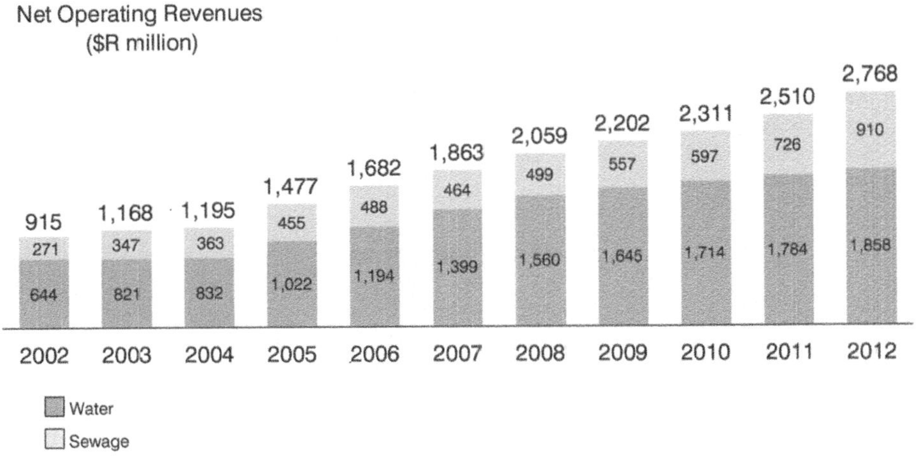


**Figure 5.3.13** Evolution of net operating revenue (2002-2012).

**Table 5.3.1** Key economic-financial indicators (2002, 2006 and 2012).

| (R$ thousands) | 2002 | 2006 | 2012 |
|---|---|---|---|
| Net Operating Income | 877,581 | 1,681,912 | 2,768,365 |
| Total Operating Costs | 775,404 | 1,256,520 | 2,005,527 |
| EBITDA | 285,000 | 656,254 | 1,145,998 |
| Net Profit | (89,950) | 356,437 | 486,928 |
| Total Assets | 2,545,402 | 4,924,649 | 8,992,688 |
| Gross Debt | 1,062,000 | 1,213,065 | 3,182,320 |
| Net Debt | 1,001,000 | 382,773 | 2,685,895 |

## 5.3.7 TECHNICAL AND OPERATIONAL PERFORMANCE

Cost reduction and efficiency shaped the evolution of the main performance indicators of *Copasa* since implementation of shock management in 2002, as outlined in Table 5.3.2.

Table 5.3.3 shows a comparison between the average operational performance of the state companies of Brazil and *Copasa*.

### 5.3.7.1 Non Revenue Water

The Non-Revenue Water (NRW) indicator was reduced to 236.1 l/connection/day in June 2013, one of the lowest values in Brazil (Figure 5.3.14).

In 2003, *Copasa* created a working group to establish and implement a Water Loss Reduction Program in line with the IWA Water Balance methodology (Souza, 2009). This program was based on four pillars, as shown in Figure 5.3.15. Evolution of NRW is outlined in Figure 5.3.16.

**Table 5.3.2** Main operational performance indicators (2002-2012).

| Indicators | 2002 | 2012 |
|---|---|---|
| Employees | 10.075 | 11,611 |
| Productivity (employees/1,000 connections) | 2.63 | 1.96 |
| Volume of Water Produced (1,000 m³/year) | 822,560 | 948,816 |
| Volume of Treated Sewage (1,000 m³/year) | 56,102 | 211,092 |
| Billed Volume (1,000 m³/year) | | |
| Water | 613,861 | 667,325 |
| Sewage | 290,502 | 428,220 |
| Total | 904,363 | 1,095,545 |
| Billing Loss Index (%) | 25.50 | 29.60 |
| Water not Converted into Revenue (l/connection/day) | 287,50 | 234,01 |

**Table 5.3.3** Operational performance: comparison between Brazil and *Copasa* (2011).

| Index | Brazil | *Copasa* |
|---|---|---|
| Population Served – Water Network (%) | 82.4 | 84.1 |
| Population Served – Sewage Network (%) | 48.1 | 65.5 |
| Water consumption per capita | 162.6 | 155.5 |
| Non-Revenue Water (%) | 38.8 | 32.5 |

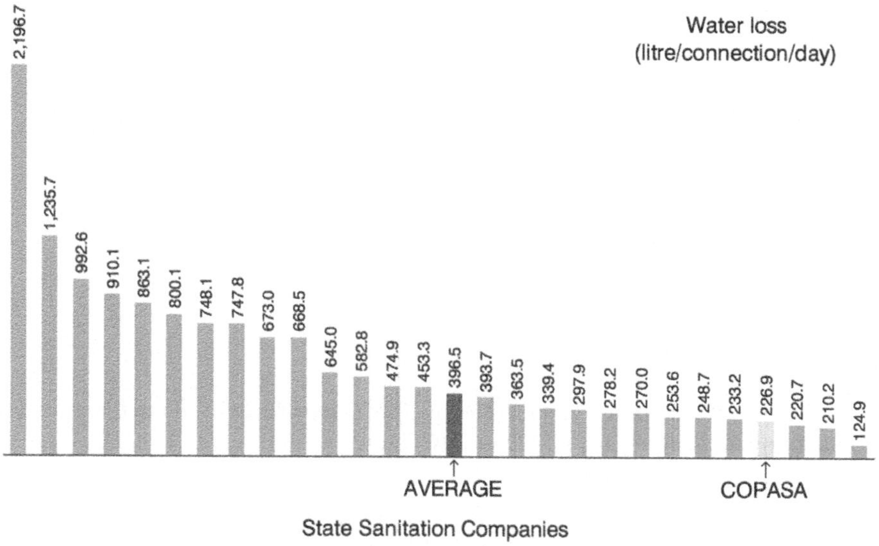


**Figure 5.3.14** Water loss rate of Brazilian companies.

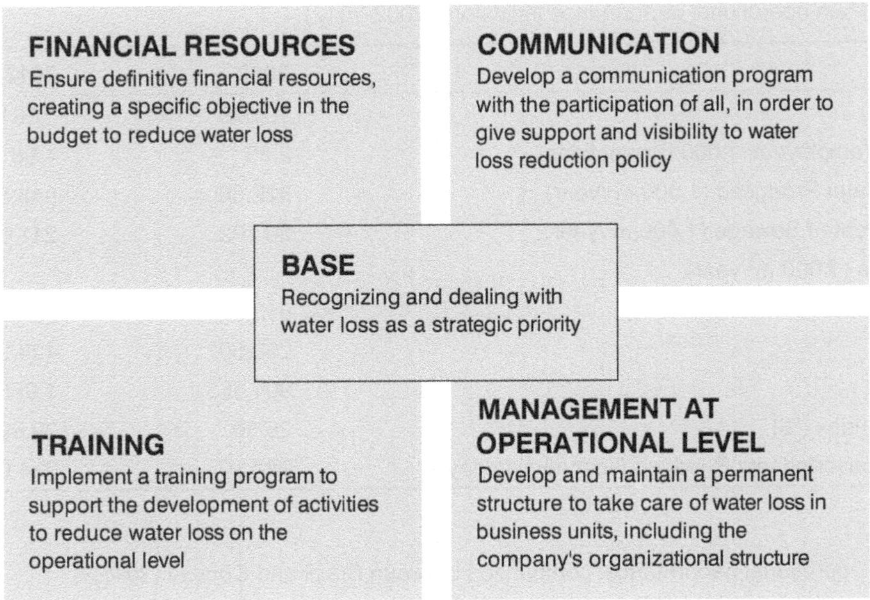


**Figure 5.3.15** Pillars of *Copasa* Water Loss Reduction Program.

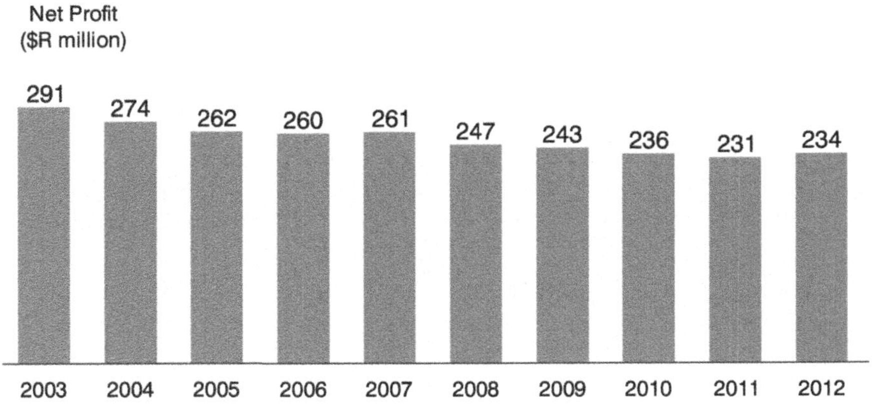


**Figure 5.3.16** Evolution of Non-Revenue Water (2003-2012).

## 5.3.7.2 Energy efficiency

Taking into account the water – energy connection, in 2009 *Copasa* approved the Integrated Program for Water Loss Reduction and Lower Cost of Electricity, known as the Energy Efficiency Program (EEP). This program assumes that water loss reduction also means lower energy consumption: the less water that is transported, along with more efficient transport of that water, the less energy that is consumed. Electricity consumption by *Copasa* in 2012 represented 12% of total costs of the company. Co-generation has been responsible for a greater reduction of energy acquired and used (Figures 5.3.17 and 5.3.18).

**Figure 5.3.17** Co-generation plant at the *Arrudas* WWTP.

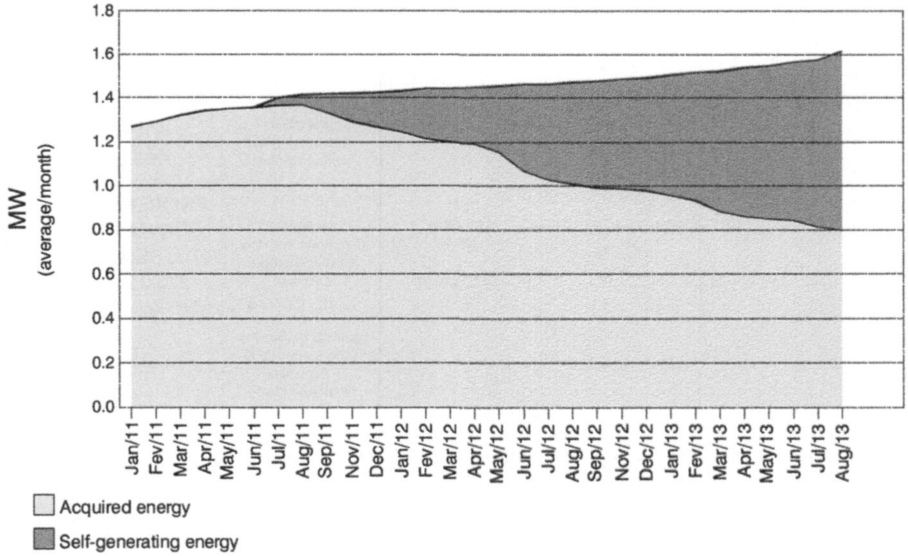


**Figure 5.3.18** Evolution of self-generated energy.

*Copasa* developed a new performance indicator, Energy Not Converted into Revenue (ENCR), that measures the quantity of energy associated with water loss reduction, taking into consideration the average energy consumption to transport each cubic metre of water (Table 5.3.4. and Figure 5.3.19).

Figure 5.3.19 illustrates the water – energy connection using the IWA Water Balance and detailing each of its components.

**Table 5.3.4** Evolution of Energy Not Converted into Revenue (ENCR) (2010-2012).

| | Consumption million kWh | ENCR (kWh/connection/day) | NRW (l/connection/day) |
|---|---|---|---|
| 2010 | 776 | 0.20 | 234 |
| 2011 | 803 | 0.19 | 231 |
| 2012 | 832 | 0.20 | 235 |

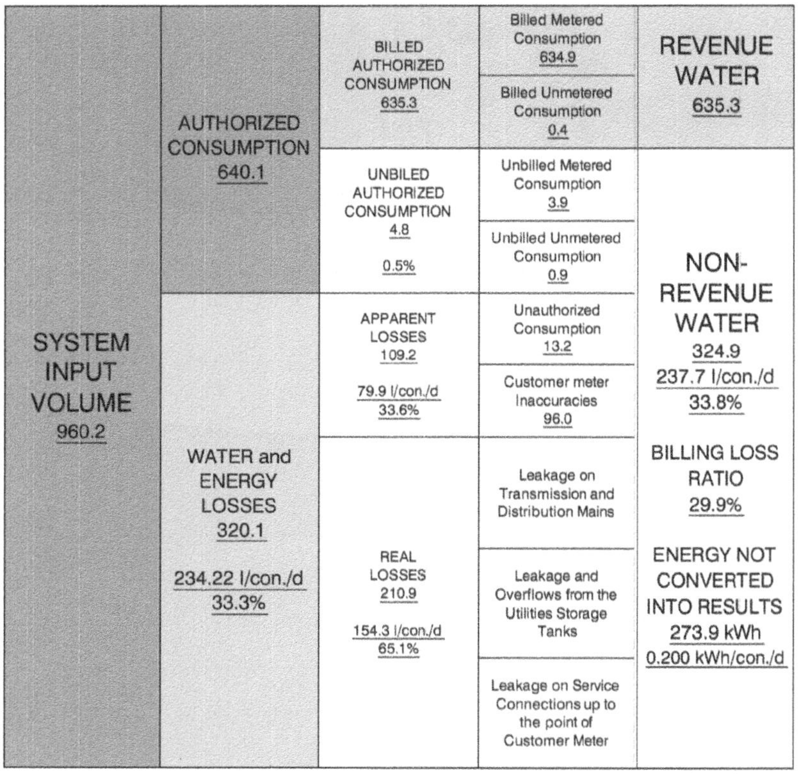


**Figure 5.3.19** Adapted IWA Water Balance: water – energy connection.

### 5.3.7.3 Water measurement indexes

The macro measurement index, namely the percentage of water produced by treatment plants that is actually measured, is 97.3% and the micro measurement index, the percentage of customers with a water meter, has reached 99.9%.

*Copasa* developed and implemented the Water Meter Park Statistical Management System (SIGEPH) consisting of a computerized system for statistical analysis of data related to customer meters. This data is stored for at least 24 months, making it possible to monitor meter age, to decide on the replacement or resizing requirements, evaluate impact of meter replacements and to detect potential frauds (Filho, 2009).

### 5.3.7.4 Index of non-payment

The overall non-payment rate, which corresponds to the ratio between the total amount not-payed and the total amount billed was 1.32% in 2012 (Figure 5.3.20). Non-payment rates have declined as a result of a consistent commercial policy, which includes development of firm actions for collection and recovery as well as a strategy of debt negotiation with major customers and government bodies (Gomes, 2009).

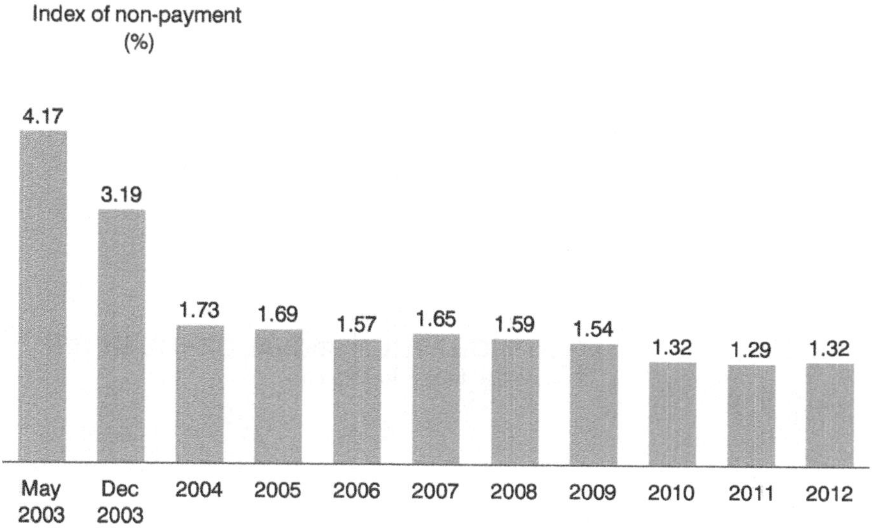


**Figure 5.3.20** Evolution of non-payment rate (2003-2012).

In 2011, *Copasa* issued debt notices for approximately 19.5% of all bills issued per month. The average water supply interruption for non-payment during that period was approximately 2.6% of all bills issued. The collection policy of water bills takes place following a number of steps illustrated in Figure 5.3.21.

## 5.3.8 COPANOR

Copanor, a Subsidiary of *Copasa*, is the concessionaire for provision of sustainable water, sewage collection and treatment services at reduced rates in 463 locations with a low Human Development Index (HDI) and populations of between 200 and 5,000 inhabitants in the northeast region of the state of *Minas Gerais*. With resources from the State Government of *Minas Gerais*, Copanor has a technological and organizational structure that allows it to increase efficiency and reduce operating costs in providing sanitation services. This enables it to maintain a tariff structure that is compatible with the socioeconomic reality of the region, without compromising service quality.

The experience of this subsidiary was one of those chosen by the World Bank as a model of implementation and management of sanitation services in isolated communities in Brazil by the "Rural Sanitation Studies and Models" project conducted by the Bank.

In December 2012, there were a total of 175 locations with a water supply system implemented, benefitting 240,000 people and 42 locations with a sewage system, benefitting 72,000 people. In addition, 930 families benefitted from the implementation of sanitary modules with a bathroom,

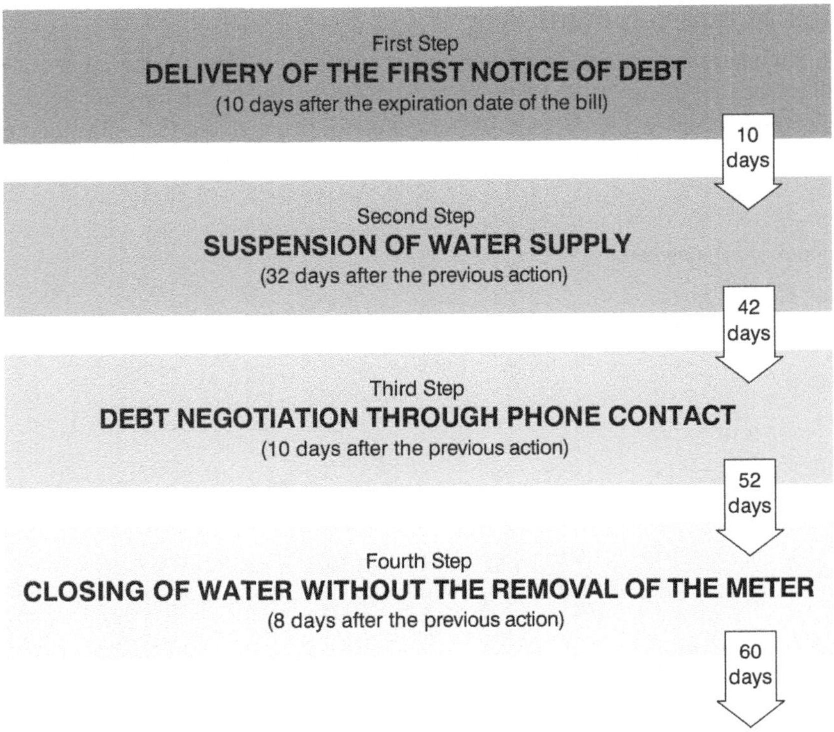


**Figure 5.3.21** Water bill collection phases.

washing tank, shower and sink in their residences. In its first stage of implementation up to 2015, 463 locations will be served with a population between 200 and 5,000 inhabitants, benefitting 400,000 people.

In 2012, R$100 million was invested in this project, mostly for drilling new wells. For 2013/14, resources were assured to the value of R$ 200 million from the State Treasury of *Minas Gerais*, with another R$ 150 million coming from BNDES/PROINVESTE.

## 5.3.9 SOCIAL RESPONSIBILITY

*Copasa* has developed a number of programs and actions, described below, in order to ensure water and sewage services to the poorest segments of the population of *Minas Gerais*.

### 5.3.9.1 Social tariff

The social tariff is a reduced tariff created in 1994 to provide more vulnerable customers with cheaper water and sewage bills. It is aimed at customers who live in properties with a building area of up to 44m² and consumption of up to 15m³ per month.

With a normative resolution approved in 2012, the regulator established new criteria for the social tariff entitlement: the customer must belong to a family enrolled in the Federal Government's Unified Social Programs Register, with a per capita income under or equal to half of the national minimum wage.

In 2012, 510,728 households (average/month) benefited from the water social tariff and 287,734 households (average/month) with sewage social tariff.

### 5.3.9.2 Progressive Discount Program

*Copasa* provides discounted rates for water and wastewater for municipalities (grantors), which are in default with the company. The discount percentage is defined as a function of default time and may reach 50% of invoice values. With up to six months of timely payment, the discount is 10%, from seven to twelve months, the discount becomes 20%, from 13 to 18 months, 30%, from 19 to 24 months it is 40% and above 24 months, 50%.

By 2011, 391 municipalities (69% of the municipal grantors) have benefited of this program and 257 received discounts of 50% in the value of the accounts.

### 5.3.9.3 "Vale Água" Program

Established in 2006, this program encourages residents from the outskirts of *Belo Horizonte* to exchange aluminium cans and plastic bottles for water bill discounts (Figure 5.3.22). The exchange of recyclable material is made in the Outpost Service of *Copasa*, through a simple procedure: the client goes to the post, weighs the material and receives a voucher. The discount is effective on the next bill. Any resident can join the program, regardless of the type of property, such as residential or commercial, churches or associations, for example.

**Figure 5.3.22** Customers exchange recyclables for discounts on their water bill.

*Copasa's* objectives with relation to the *"Vale Água"* Program areas are to:

- Benefit its low-income customers by promoting a reduction in the value of the water bill;
- Ensure access to basic sanitation services to a growing population;
- Create environmental awareness of recycling;
- Reduce the volume of waste released into water courses and the environment;
- Reduce flooding and the incidence of waterborne diseases.

The program is present in six areas around *Belo Horizonte* and, in 2011, received 3,500 kg of aluminium cans and 40,000 kg of plastic bottles.

#### 5.3.9.4 Subsidy to Philanthropic Entities

*Copasa* can allocate up to 0.6% of its monthly earnings to subsidies, held in collaboration with the State Department of Social Welfare and Voluntary Social Service Assistance. In 2012, 748 charities have benefited from discounted rates for water and sewer services up to the limit of contracted demand.

### 5.3.10 CREDITS AND ACKNOWLEDGEMENTS

The creators of the shock management concept in *Minas Gerais* were the former Governor Aécio Neves and the former Vice-Governor and later Governor Antônio Anastasia.

Mauro Ricardo Costa was the first CEO of *Copasa* and implemented the shock management in the company.

Ricardo Augusto Simões Campos, as CFO of *Copasa*, was responsible for the listing of the company at BOVESPA and later, as CEO, pursued the change management process and chaired the company towards stability.

### 5.3.11 REFERENCES

Anastasia A. A. (2009). Municipal Governance: Shock Management in *Minas Gerais*. V State Seminar on Innovation in Public Management, *Minas Gerais*, Brazil, August 2009.

*Copasa* (2009). Annual Report and Sustainability Report 2008. *Copasa*, *Minas Gerais*, Brazil.

*Copasa* (2013). Annual Report and Sustainability Report 2012. *Copasa*, *Minas Gerais*, Brazil.

Filho A. R. C. (2009). Hydrometry Division of *Copasa*. *Copasa*, *Belo Horizonte*, Brazil.

Gomes C. (2009). Non-Payment in *Copasa*: Commercial Policies and Programs. *Copasa*, *Belo Horizonte*, Brazil.

Souza P. C. (2009). Best Management Practices Associated with Technological Solutions Adopted in *Copasa* to Reduce Water Loss. *Copasa, Belo Horizonte*, Brazil.

# Chapter 5.4

# Case study 4 - PNQS towards excellence

*With Cassilda Teixeira de Carvalho*
*President of the PNQS and former President of ABES*

## 5.4.1 INTRODUCTION

It is clear that there is a water crisis in the world, although this crisis is not so much about lack of water but rather more about the lack of capability to manage it. Considering this issue, ABES, the Brazilian Association of Sanitary and Environmental Engineering, has since 1997 developed and managed a unique national contest, the PNQS – National Quality Sanitation Award. The PNQS is geared towards improving otherwise missing management capability in Brazil through a model which can be used in many other countries around the world. This is really much more than just an award; it is a powerful change management tool that has led and continues to lead water companies towards excellence.

To participate in the PNQS, every year thousands of employees from dozens of Brazilian water companies take part in specialized training courses and workshops on technical and management subjects. The PNQS mission statement is the stimulation of adoption of worldwide best practices by Brazilian companies through the identification, recognition and sharing of success management cases, which may contribute to water sector improvement and to people's quality of life.

The PNQS vision statement is becoming a reference tool to appraise the management system and results of Brazilian water companies. It is based on four pillars: (1) mobilization, which usually involves large teams in each water company; (2) training by participation in management courses at various levels; (3) evaluation with company management systems and practices being evaluated according to the same criteria, according to their maturity levels; and (4) recognition through awarding prizes at a national ceremony followed by a working visit of the winner's managers to some of the best water companies in the world.

There are several award categories. A water company or a business unit may apply for an award in five different categories, corresponding to maturity levels: 125, 250, 500, 750 and 1000 points. In addition, the IGS – Innovation in Sanitation Management – is a special category geared towards revealing centres of excellence which may exist in any organization, even in those with a low maturity level. A great effort is made to ensure the credibility of the whole process, for instance, by having a clear regulation, common evaluation criteria and a trained board of independent examiners.

Those who have been following the evolution of the sanitary engineering sector in Brazil, especially after the creation in 1968 of the *Sistema Financeiro de Saneamento* (SFS), followed by the *Banco Nacional da Habitação* (BNH) and the *Plano Nacional de Saneamento* (Planasa) in 1971, have seen that great progress has been made since then.

As a concept, the PNQS belongs to ABES, but as a product it belongs to all entities that operate in Brazil's water sector. The sector as a whole is currently revisiting the problems that are truly essential in Brazil and, indeed, in every country of the world – innovation, new technologies, service engineering, quality, management and products. The environmental issue, amplified by the media, has generated a collective conscience - local and global - about quality of life and citizens' rights. More and more, the people express their concerns and demands in various associations and in public hearings.

An optimistic indicator revealed by PNQS is the maturity level that the sanitation sector has reached. Replacing the old debate on institutional competences, the issue is now a search for quality management and excellence at all levels, always looking to achieve the best results for customers. In Brazil, municipalities may choose to manage services directly, to transfer them to the state company or to grant them to a private company. The local reality varies from one company to another and indeed only the municipal authority is able to decide the best option to take. Even when a municipality signs a concession contract with other entities, it does not alienate its ultimate global responsibility for the water or wastewater services.

A private sector initiative has stimulated a healthy competition in the water sector. The arrival of the first international private company occurred in 1994, in the municipality of Limeira.

### 5.4.2 ORIGIN AND CONCEPTION OF PNQS

The US Secretary of Commerce was responsible for developing a recovery program for the industrial sector in the early 1980s. The first option of this program concerned excellence in management, to achieve quality of products and international competitiveness. Another example of success to be followed in this area is the Japanese *Total Quality Management* program.

Awards and excellence models in management spread quickly all over the world, for instance the European Foundation of Quality, the Singapore Quality Award, the South Africa Quality Award and the Australia Quality Award. In 2013, there were some 75 prizes of this type in over one hundred countries.

In Brazil, the first national quality award system was introduced in 1991 by the National Committee for Quality and Productivity, under the responsibility of the minister of Industry, Trade and Tourism. The project, sponsored by private and public companies, created the National Quality Award Foundation (FPNQ, now FNQ) to reproduce a Brazilian version of the American Malcolm Baldrige Award in 1992 (Figure 5.4.1).

The first four editions of the National Quality Award (PNQ) were won by multinational companies installed in Brazil, already familiar with quality management. In 1995-96 the Gerdau Foundation launched a program to stimulate quality management in the Brazilian state of Rio Grande do Sul. Inspired by those strategies, in 1997 ABES created the ABES Quality Award program, renamed in 2000 as PNQS – National Sanitation Quality Award (Figure 5.4.2). The program was designed to seek and share successful local and international management experiences in order to develop the

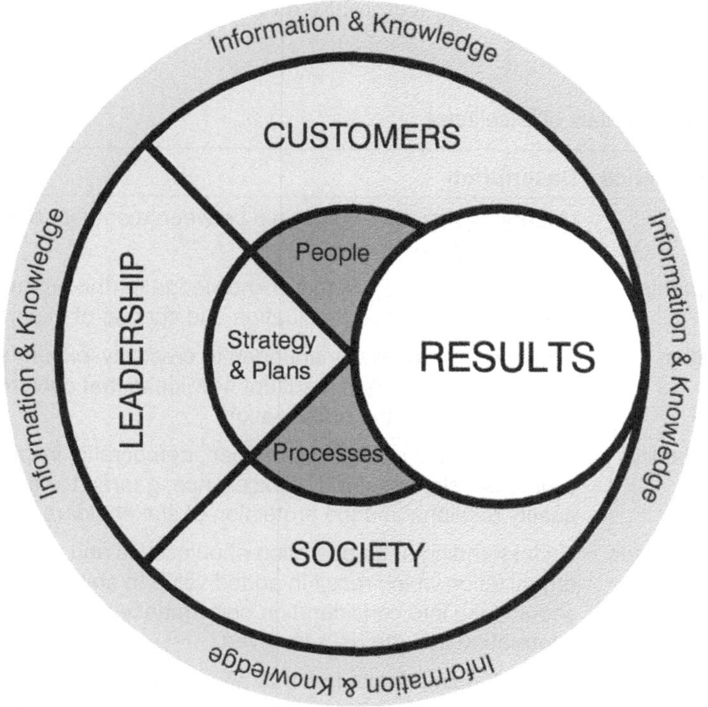


**Figure 5.4.1** The FNQ-National Quality Foundation MEG® (Management Excellence Model) diagram (Source: ABES PNQS, 2000).

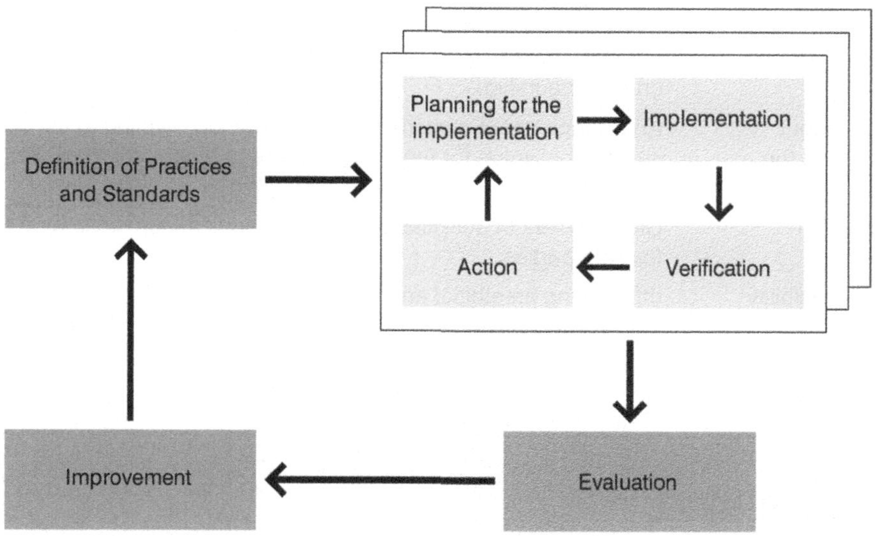


**Figure 5.4.2** The PNQS action plan (Source: ABES PNQS, 2000).

Brazilian water, wastewater and solid waste sector, based on the Fundaments of Excellence presented in Table 5.4.1.

**Table 5.4.1** PNQS fundamentals of excellence.

| Fundamentals of excellence | Description |
|---|---|
| 1. Systemic Thought | Understanding the relationship between members of the organization and external environment |
| 2. Organizational Learning | Searching for a new stage of knowledge for the organization through the perception, reflection, evaluation and sharing of experiences |
| 3. Culture of Innovation | Creation of an environment open to creativity, experimentation and implementation of resultant new ideas that can create positive differentiation for the organization |
| 4. Leadership and Stability of Targets | Leaders should perform in an open, democratic, inspiring and motivating way to people, in search of excellence, geared towards the promotion of quality relations and the protection of stakeholders |
| 5. Orientation for Processes and Information | Understanding and separation of objectives and processes of the organization which result in added value to stakeholders; decisions should take into consideration performance evaluation, the available information and the risks involved. |
| 6. Vision of the Future | Understanding the factors that affect the organization, its ecosystem and the external environment in the short and long run, towards sustainability. |
| 7. Value Generation | Achieving of results consistent with the sustainability of the organization through the increase of tangible and intangible value for the stakeholders. |
| 8. Development of People | Establishment of relations with people leading to professional and human fulfilment, improving their performance through commitment, development of competences and opportunities for personal entrepreneurship. |
| 9. Knowledge about Customers and the Market | Knowledge about customers and the market, leading to the creation of sustainable value for the customer and to higher competitiveness in the markets. |
| 10. Development of Partnerships | Development of activities with other organizations, searching for benefits for both parties. |
| 11. Social Responsibility | Introducing an ethical and transparent relation with all the stakeholders in the organization strategy, geared towards the sustainable development of society, preserving the environmental and cultural resources for future generations, respecting differences and promoting the reduction of social asymmetries. |

## 5.4.3 POLITICAL SKILL

As a civil society entity, ABES recognizes that it should not go beyond its technical and institutional boundaries and attempt to replace governments or companies or to develop official policies. The main philosophy of PNQS started from the belief that it would not be for ABES to introduce a business excellence program or to play a consultant role in affiliated companies. The idea was to remain equidistant, neutral and impartial in order to be accepted as a know-how centre, a state of the art repository of best management and technical practices, working with its members and other stakeholders.

The PNQS is an award, a stimulus and an incentive to companies that are already willing to develop their own business excellence programs. It helps to scale up the results of their own programs. Moreover, one of the novelties of the award is the publication and dissemination in the Brazilian companies of a common and comprehensive Indicators Guide (described below). The competition is endogenous, namely in that candidates who undertake the award are judged according to the evaluation of the results achieved in their projects of excellence in accordance with the annual regulation of the National Committee of Quality ABES (CNQA). In other words, the company competes with itself.

The PNQS does not prescribe any specific practice or method. It just evaluates what is working successfully and producing results in the applicant water companies and shares it within the whole sector. Regulation is quite extensive and detailed, yet without being too complex, in order to enable many candidates to apply.

When a company applies for the award, it knows the evaluation criteria in advance. It is obvious that a company is not a candidate if it is not able to fulfil the regulation requirements. At the same time, to be successful, a company must develop its own excellence program, no matter where the inspiration comes from: local, American, European, Japanese or any other. ABES does not impose any existing commercial package. It only prepares a regulation that contains a number of assumptions that are listed in any excellence program. Even if a company did not contract external consultants, it may compete for the award with the work, dedication, belief and enthusiasm of its own employees.

The main characteristic of the PNQS tool – and the PNQS is nothing more than a tool – is its flexibility. This tool, adapted to the sanitation sector, will serves to evaluate the management system and its results either in an individual business unit or in an entire company. This means that the PNQS tool is also elastic: it can grow or shrink according to the target. The more advanced the company business excellence program is, the more it will be encouraged to apply for the highest levels of PNQS. If the company is starting to take its first steps towards excellence, it is advisable that it is a candidate to Level "B" (Basic). As advances are shown and depending on commitment with excellence, it is possible to ascend to higher levels.

It is interesting that many companies are introducing internal awarding systems as multipliers of the PNQS concepts. ABES has provided a large amount of information and material about PNQS to companies and professionals in the sector, namely manuals, courses and workshops.

## 5.4.4 DESIGN, STRUCTURING AND EVOLUTION OF PNQS

Created in 1997, the PNQS launched its first cycle with just one level of 250 evaluation points and had the intention of being biannual. In the following year, as a result of its enormous acceptance and the requests of the most diverse industry segments, PNQS became annual. In its third year of existence, PNQ the S launched Level II with 500 points, followed by Level III with 750 points and finally Level IV with 1000 points. The 125 point model was introduced in 2011 to attract beginners.

The idea of creating the PNQS was not original, as the program was inspired by the best programs throughout the world. The task of ABES was to adjust procedures and evaluation factors to the specific realities of the sanitation sector in Brazil.

The management of PNQS is undertaken by the National Committee of Quality of ABES (CNQA). The CNQA manages all steps of PNQS including the regulation of the prizes, evaluation process, selection and training of examiners and judges and training courses on management development.

The PNQS is intended for public or private companies, state or municipal, with operations in Brazil, in the various segments of the environmental sanitation: water supply, wastewater, stormwater and solid wastes.

Application to PNQS is a first step towards the recognition of the level of maturity that a company has achieved. The award encourages companies to climb the various PNQS maturity levels. Eligibility for the award is made based on the evaluation of the management systems in all the eight components of PNQS as listed in Table 5.4.2.

**Table 5.4.2** The eight components of PNQS.

| PNQS Component | Description |
|---|---|
| 1. Leadership | Corporate governance; leadership and promotion of an excellence culture; analysis of organization performance. |
| 2. Strategies and plans | Development of strategies and implementation of those strategies. |
| 3. Clients | Relationship with clients; image and marketing. |
| 4. Society | Socio-environmental responsibility and social development. |
| 5. Information and knowledge | Organization information; intangible assets and organizational knowledge. |
| 6. People | Working systems; training and development; quality of life. |
| 7. Processes | Core business and supporting processes; supplier related processes; economic and financial processes. |
| 8. Results | Financial; regarding customers and the market, people, society, processes and suppliers. |

The maximum classifications corresponding to each criteria and sub-criteria are presented in Table 5.4.3.

In 2000, PNQS introduced a Reference Guide for Performance Measurement (GRMD) adapted to the reality of the sanitation sector in Brazil, including the awards structure (CNQA, 2013). This guide defines indicators, presented in the supplementary information, that lead to the definition of a company's global level of maturity based on a number of different perspectives (Table 5.4.4). The advantages of the GRMD are twofold: on one hand, it helps build a valuable balanced scorecard whilst on the other hand, it allows for benchmarking in the sector.

Since its creation, it has been the duty of ABES to promote and stimulate development and training of human resources for the environmental sanitation sector. The training courses are now a pre-requirement for applying to PNQS and they were based on the assumption that all participants would be committed to improve their management skills, continuously seeking alignment with the PNQS evaluation criteria. The Innovation in Sanitation Management (IGS) category, introduced in 2004, captures particular management experiences and their results through special evaluation criteria based on business excellence principles (ABES PNQS, 2000).

IGS is particularly recommended for an organization that is not considered fully prepared to compete in the other categories, but has already obtained proven success in some management process. In fact,

**Table 5.4.3** Maximum classifications.

| | **Criteria** | **LEVELS** | | | | |
|---|---|---|---|---|---|---|
| | | **B** | **I** | **II** | **III** | **IV** |
| Leadership | **TOTAL** | **13** | **25** | **50** | **75** | **100** |
| | Corporate governance | | | 50 | 75 | 100 |
| | Leadership and culture of excellence | | | 15 | 25 | 30 |
| | Analysis of organization performance | | | 20 | 25 | 40 |
| Strategies and Plans | **TOTAL** | **13** | **25** | **40** | **60** | **80** |
| | Strategy formulation | | | 20 | 25 | 40 |
| | Strategy implementation | | | 20 | 35 | 40 |
| Customers | **TOTAL** | **13** | **25** | **30** | **45** | **60** |
| | Image and knowledge of the market | | | 10 | 20 | 20 |
| | Customer Relations | | | 20 | 25 | 40 |
| Society | **TOTAL** | **10** | **20** | **30** | **45** | **60** |
| | Environmental and social responsibility | | | 15 | 30 | 30 |
| | Social development | | | 15 | 15 | 30 |
| Information and Knowledge | **TOTAL** | **10** | **20** | **25** | **35** | **50** |
| | Organizational information | | | 15 | 20 | 30 |
| | Organizational knowledge | | | 10 | 15 | 20 |
| People | **TOTAL** | **10** | **20** | **40** | **60** | **80** |
| | Working systems | | | 30 | 40 | 60 |
| | Training and development | | | 15 | 20 | 30 |
| | Quality of life | | | 10 | 15 | 20 |
| Processes | **TOTAL** | **13** | **25** | **55** | **85** | **110** |
| | Main business and supporting processes | | | 30 | 40 | 60 |
| | Suppliers processes | | | 15 | 25 | 30 |
| | Economic/financial processes | | | 10 | 20 | 20 |
| Results | **TOTAL** | **43** | **90** | **230** | **345** | **460** |
| | Economic/financial | | | 50 | 75 | 100 |
| | Relative to customers and the market | | | 50 | 75 | 100 |
| | Relative to Society | | | 30 | 45 | 60 |
| | Relative to People | | | 30 | 45 | 60 |
| | Relative to Processes | | | 50 | 75 | 100 |
| | Relative to Suppliers | | | 20 | 30 | 40 |
| **Points** | | **125** | **250** | **500** | **750** | **1000** |

**Table 5.4.4** Levels of maturity of water utilities.

| LEVEL OF MATURITY | Description | POINTS |
|---|---|---|
| 9 | Highly pro-active, refined, innovative and continuously updated objectives; leadership in the sector widely acknowledged as a reference of excellence in all areas of activity, processes and products; generalized improvement of all the results; the stakeholders issues are correctly addressed. | 851 - 1000 |
| 8 | Significant pro-active, innovative and updated objectives; recognition in the sector in some areas of activity, processes and products; stakeholders issues are correctly addressed. | 751 - 850 |
| 7 | Most objectives supported on training and innovation; pro-activity is the rule; still some gaps of interaction between activity areas. | 651 - 750 |
| 6 | Adequate objectives defined for majority of issues; virtually full alignment between corporate strategies and management actions; extended training and innovation practices; stakeholders issues addressed correctly. | 551 - 650 |
| 5 | Generalized alignment between corporate strategies and management actions; early results from training and innovation already achieved; most stakeholder issues are correctly addressed. | 451 - 550 |
| 4 | Generalized implementation of proactive procedures; significant alignment between corporate strategies and management actions in most situations; many of the issues related to stakeholders are correctly addressed. | 351 - 450 |
| 3 | Wide implementation of proactive procedures; still lack of coherence between corporate strategies and management actions; significant attention given to stakeholders. | 251 - 350 |
| 2 | Initial stages of implementation of proactive procedures and integration of processes; some attention to stakeholders. | 151 - 250 |
| 1 | Preliminary stage of development of a water utility; mostly reactive procedures and individual learning; virtually no innovation; limited attention to stakeholders | 0 - 150 |

in almost every organization it is possible to find some management centres of excellence. IGS is an opportunity to ensure that they are acknowledged and disclosed to the sector. IGS also serves as a platform to inoculate business excellence concepts in organizations.

As the PNQS purpose is to reward skills, it is normal to express this recognition through prizes: trophies, medals, plaques and diplomas. ABES believes that the gala awarding event is a very solemn occasion. Organized annually by CNQA, the Quiron trophy has become the "Oscar" of the environmental sanitation sector in Brazil, due to its symbolism in the sanitation sector. There are Quiron Diamond, Platinum, Gold, Silver, Bronze and Copper trophies to the winners of Levels IV, III, II, I and B, respectively. Two IGS category winners are awarded with the "Innovation in Sanitation Management" Medal. The other finalist cases, around 10 in total, are rewarded with the "Finalist in Innovation in Sanitation Management" Plate.

The PNQS Level B to Level IV award-winners and two IGS winners are selected to participate in a mission abroad (Mission of Outdoor Studies). The international PNQS Study Mission crowns this entire process and has the objective to exchange experiences in the field of environmental sanitation. Each year, a specific schedule covers a set of organizations of other countries, indicated and selected based on topics of relevance to the sanitation sector in Brazil, such as excellence in control of water quality, the integrated management of water resources and the innovative technologies applied to treatment effluents, among others. Also known as the Benchmark Mission, the visits have already taken hundreds of professionals to exchange knowledge in countries like France, Germany, England, Canada, United States, Spain, Italy, Portugal, Holland and Belgium. The PNQS missions encompass a virtuous cycle. The countries to be visited are chosen by their accumulated experiences and results measured in processes of engineering, technology and management. Each trip is carefully prepared and executed with guided visits to selected installations and technical seminars which guarantee participants the best advantage in terms of knowledge and innovation that they internalize and share with their working peers.

## 5.4.5 CONCLUSIONS

PNQS cannot replace the government spheres in public policies competence for sanitation. Instead, it is a tool that helps to qualify both companies and professionals in the sector. The innovation and management models applied to water companies certainly result in more satisfied customers and better quality of service and products. Several indicators point in that direction, for example the reduction in the national average of water losses in recent years from 48% to 38%.

However, at this stage of the development of Brazil, the greater ABES commitment, as a civil society entity, is with the provision of environmental sanitation services for all the Brazilian people, in all regions and all social segments. The current figures demonstrate the impact of PNQS in the Brazilian sanitation sector: 472 applications, more than 12,000 managers trained in 22 states, 230 organizations recognized with the PNQS and 106 IGS finalists.

## 5.4.6 REFERENCES

ABES PNQS (2000). Reference Guide for Performance Measurement (GRMD). ABES, Rio de Janeiro, Brazil.
CNQA National Quality Committee ABES (2013). Guide PNQS 2013. ABES, Rio de Janeiro, Brazil.

## *Supplementary Information*

Indicators used in PNQS
(fully described in the PNQS Regulation)

**Appendix:** Indicators used in PNQS (fully described in the PNQS Regulation).

| **Economic and Financial Indicators** | **Code** | **B** | **I** | **II** | **III** | **IV** |
|---|---|---|---|---|---|---|
| Financial performance indicator | IFn01 | No | No | No | No | No |
| Volume of Non Revenue Water | IFn02 | Yes | No | No | No | No |
| Total cost of the services /m$^3$ | IFn03 | Yes | No | No | No | No |
| Execution of the budget investment | IFn04 | Yes | Yes | No | No | No |
| Liquid margin with depreciation | IFn05 | Yes | Yes | No | No | No |
| Total investment | IFn06 | Yes | Yes | Yes | No | No |
| Number of days of revenue to cover the customer debt | IFn07 | Yes | No | No | No | No |
| Cost/capita with handling of municipal solid wastes | IFn08 | No | No | No | No | No |
| Cost recovery of stormwater collection | IFn09 | No | No | No | No | No |
| Cost recovery of handling of municipal solid wastes | IFn10 | No | No | No | No | No |
| Cost/capita with stormwater collection | IFn11 | No | No | No | No | No |
| Solid wastes tax/building | IFn12 | Yes | Yes | Yes | Yes | No |
| Stormwater tax/building | IFn13 | Yes | Yes | Yes | Yes | No |
| Non-payment indicator | IFn15 | No | No | No | No | No |
| Rate of return on Liquid Assets | IFn16 | Yes | Yes | Yes | Yes | Yes |

| **Customers and market related indicators** | **Code** | **B** | **I** | **II** | **III** | **IV** |
|---|---|---|---|---|---|---|
| Complaints and communication of problems indicator | ICm01 | Yes | No | No | No | No |
| Complaints indicator | ICm01a | Yes | No | No | No | No |
| Customer satisfaction indicator | ICM02 | No | No | No | No | No |
| Organization image satisfaction indicator | ICM03 | Yes | Yes | No | No | No |
| Awareness of services and products indicator | ICM04 | Yes | Yes | No | No | No |
| Urban water supply coverage indicator | ICM05 | No | Yes | Yes | No | No |
| Urban wastewater collection coverage indicator | ICM06 | No | No | No | No | No |
| Wastewater treatment coverage indicator | ICM07 | Yes | No | No | No | No |
| Total water supply coverage indicator | ICM08 | Yes | Yes | Yes | Yes | No |
| Total wastewater coverage indicator | ICM09 | Yes | Yes | Yes | Yes | No |
| Average response time to customer complaints | ICM10 | Yes | No | No | No | No |
| Urban solid wastes coverage indicator | ICM11 | No | No | No | No | No |
| Urban stormwater coverage indicator | ICM12 | No | No | No | No | No |
| Complaints addressed to consumers authority | ICM13 | Yes | Yes | Yes | Yes | No |
| Hearings in the consumers authority | ICM14 | Yes | Yes | Yes | Yes | No |

*(Continued)*

**Appendix:** Indicators used in PNQS (fully described in the PNQS Regulation) (*Continued*).

| Society related indicators | Code | B | I | II | III | IV |
|---|---|---|---|---|---|---|
| Social development indicator | ISc01 | Yes | Yes | Yes | Yes | No |
| Sanctions and indemnities indicator | ISc02 | No | No | No | No | No |
| Solid wastes sanctions and indemnities indicator | ISc02a | No | No | No | No | No |
| Mitigation of environmental impacts indicator | ISc04 | Yes | No | No | No | No |
| Coordination in administrative processes indicator | ISc05 | Yes | Yes | Yes | Yes | No |
| Coordination in judicial processes indicator | ISc07 | Yes | Yes | No | No | No |
| Family income indicator | ISc08 | Yes | Yes | Yes | Yes | No |
| Wastewater treatment indicator | ISc09 | No | No | No | No | No |

| People related indicators | Code | B | I | II | III | IV |
|---|---|---|---|---|---|---|
| Productivity of the labour force indicator | Ipe01 | Yes | No | No | No | No |
| Creativity of the labour force indicator | Ipe02 | Yes | No | No | No | No |
| Annual labour force training indicator | Ipe03 | No | No | Yes | Yes | No |
| Accomplishment of training plan indicator | Ipe3a | Yes | Yes | No | No | No |
| Effectiveness of training indicator | Ipe3b | Yes | Yes | No | No | No |
| Satisfaction of employees indicator | Ipe04 | No | No | No | No | No |
| Frequency of accidents indicator | Ipe05 | No | No | No | No | No |
| Seriousness of accidents indicator | Ipe06 | Yes | No | No | No | No |
| Productivity of solid wastes collection labour force indicator | Ipe07 | Yes | No | No | No | No |
| Productivity of stormwater labour force indicator | Ipe08 | Yes | No | No | No | No |
| Commitment of labour force to management directives indicator | Ipe09 | Yes | Yes | No | No | No |
| Global productivity indicator | Ipe11 | Yes | No | No | No | No |
| Absenteeism indicator | Ipe12 | Yes | Yes | Yes | No | No |

| Process related indicators | Code | B | I | II | III | IV |
|---|---|---|---|---|---|---|
| Conformity of quantity of water quality samples indicator | ISp01 | Yes | No | No | No | No |
| Water quality samples outside the standard indicator | ISp02 | No | No | No | No | No |
| Removal of pollution in the WWTP indicator | ISp03 | No | No | No | No | No |
| Average time to establish a water supply connection indicator | ISp04 | Yes | No | No | No | No |
| Average time to establish a waste water connection indicator | ISp06 | Yes | No | No | No | No |
| Complaints on water quality indicator | ISp08 | Yes | No | No | No | No |
| Complaints on interruptions of water supply indicator | ISp09 | Yes | No | No | No | No |
| Average time to deliver the services indicator | ISp10 | No | No | No | No | No |

(*Continued*)

**Appendix:** Indicators used in PNQS (fully described in the PNQS Regulation) (*Continued*).

| Process related indicators | Code | B | I | II | III | IV |
|---|---|---|---|---|---|---|
| Continuity of water supply indicator | ISp11 | Yes | No | No | No | No |
| Overflows from the wastewater system indicator | ISp13 | Yes | No | No | No | No |
| Water loss per connection indicator | ISp14 | No | No | No | No | No |
| Non conformity of residual chlorine analysis indicator | ISp15 | Yes | No | No | No | No |
| Non conformity of turbidity analysis indicator | ISp16 | Yes | No | No | No | No |
| Non conformity of total coliform analysis indicator | ISp17 | Yes | No | No | No | No |
| Conformity of quantity of water quality samples indicator | ISp18 | Yes | Yes | Yes | Yes | No |
| Reduction of the quantity of sewage collected in the system indicator | ISp19 | Yes | No | No | No | No |
| Non conformity of treated water quality samples indicator | ISp20 | Yes | No | No | No | No |
| Recovery of recyclable materials indicator | ISp21 | Yes | No | No | No | No |
| Solid wastes with adequate ultimate disposal indicator | ISp22 | No | No | No | No | No |
| Quantity of solid waste samples indicator | ISp23 | Yes | Yes | Yes | Yes | Yes |
| Delays of payment to suppliers indicator | IPa01 | Yes | No | No | No | No |
| Customer water metering indicator | IPa02 | No | No | No | No | No |
| Macro water metering indicator | IPa03 | No | No | No | No | No |
| Average consumption of electric energy indicator | IPa04 | Yes | No | No | No | No |
| Pro-active repairs indicator | IPa05 | Yes | Yes | Yes | Yes | No |
| Strategic projects implemented on time indicator | IPa07 | Yes | Yes | No | No | No |
| External evaluation of the management system indicator | IPa08 | Yes | Yes | Yes | No | No |
| Satisfaction of users of the information supplied indicator | IPa09 | Yes | No | No | No | No |
| External services completed on time indicator | IPa10 | Yes | Yes | Yes | Yes | No |

| Suppliers related indicators | Code | B | I | II | III | IV |
|---|---|---|---|---|---|---|
| Deliveries from suppliers indicator | IFr01 | Yes | No | No | No | No |
| Deliveries of chemical products not meeting the specification indicator | IFR02 | Yes | No | No | No | No |
| Delivers of non chemical products not meeting the specification indicator | IFr02a | Yes | Yes | Yes | Yes | No |
| Interruption of electric energy supply indicator | ISp03 | Yes | Yes | No | No | No |
| Water treatment cost indicator (indirectly measures the quality of the raw water) | ISp06 | Yes | Yes | Yes | No | No |
| Supplier performance indicator | ISp07 | No | No | No | No | No |

# Chapter 5.5

# Case study 5 - *Águas de Portugal* (2004-2005)

*Joaquim Poças Martins*
*CEO (2004–2005)*

## 5.5.1 INTRODUCTION

In 2004 a Portuguese Government Resolution was made, regarding the restructuring of the water utilities sector, with apparently clear objectives and deadlines. There was political support for the project; the shareholders appointed a new CEO to lead it; the international and national economic conditions were favourable; the access to credit with good conditions was relatively easy and there were significant European structural funds available for the three year mandate ahead. Therefore, the textbook requirements for a successful change management process appeared to have been met.

However, the political situation was completely altered within one year and the process aborted because the new Government did not support the resolution approved by the previous one. In addition, there was a major political issue in the resolution that had not been clearly addressed and could affect the whole process: namely the autonomy of the municipalities. In fact, the municipalities had responsibility for their municipal water utilities and the resolution did not take a political position on their future role in the water sector. Indeed, it implicitly assumed that this critical aspect would be resolved during the restructuring process, but without creating the necessary political conditions for that to happen.

This is one of many examples of a change management project that has ceased much earlier than anticipated due to an unforeseen interruption of a political cycle but which also did not have clearly defined objectives from inception. People often stay that the devil is in the detail: pushing the difficult political problems ahead and making decisions or anticipating investments that would only make sense if they had been resolved, may have been a serious error of judgement.

This case study confirms the ring chain change management model presented in Chapter 2, namely the political support requirement. It is an example of an attempt to reach universal coverage and sustainability at national level that could be useful elsewhere, if conveniently adapted. Similar to the Copanor experience described in the Copasa case study, the state decided to take an important step towards the universal coverage of the population with water supply and wastewater systems.

There was a major difference, though: in the Portuguese case, the government issued a general resolution stating the objectives but did not guarantee all of the indispensible success conditions from the beginning of the process.

The Copanor experience was more limited in scope but more realistic: it was aimed at a specific low income population, in a certain geographical zone of the state. It was assumed as an option for improved systems instead of completely new urban type systems. State funds were provided to finance the project and Copasa was given responsibility by the state to develop the project, without affecting its annual results and tariffs and without charging the existing Copasa customers.

## 5.5.2 COMPANY PROFILE AND BACKGROUND

*Águas de Portugal* (ADP) was, in 2004, a 100% public national holding of 67 companies created to support a dual model for water supply, wastewater and solid waste in the country (*Águas de Portugal*, 2005). At the time, it served around half of the Portuguese population in terms of bulk water supply and one third in terms of wastewater treatment.

According to the government resolution, ADP was the main vehicle for sector restructuring in the country, with the following major objectives:

1. Restructuring ADP, namely selling the international and distribution companies in competition with the private sector;
2. Universal coverage of the population, maintaining existing responsibilities of the municipalities for their systems;
3. Within five years (by 2008) completion of the infrastructure cycle of bulk water supply, wastewater and solid wastes systems in the country, where water distribution, wastewater and solid wastes collection were mostly under the responsibility of municipalities;
4. Promoting compatibility between bulk and municipal systems;
5. Promoting sustainable solutions and regional solidarity.

The forecasted investment plan of ADP in bulk water supply and wastewater infrastructure at the time amounted to 3,500 million Euros (about €350/inhabitant) for the 2004-2010 period. A similar investment plan was to be undertaken by the municipalities, in approximately the same period, to complete water supply and wastewater systems.

However, unlike the ADP investment plan, that of the municipalities' as only indicative. It was envisaged that an important part of it could be obtained from European structural funds, but it was assumed a significant self-financing component would be found through tariffs from the municipal systems. These tariffs varied greatly between the municipalities, even in adjacent ones served by the same bulk water supply or wastewater system. The majority of the systems directly managed by the municipalities did not charge sustainable tariffs to their customers and indeed, in the more rural inland areas, where the capital and operational costs were higher, tariffs were much lower. Additionally, even without reliably audited information, many of the municipal systems were believed to be run at relatively low efficiency levels, especially in terms of NRW.

## 5.5.3 THE CHALLENGES

In addition to resolving internal problems of ADP, such as the potential lack of sustainability of some bulk water companies, and selling the international and internal companies which were in competition with the private sector, there were two main challenges at the time: (1) to expand the existing water supply and wastewater systems; and (2) to reach a global sustainable solution at a national level, in the context of a dual model (bulk/retail) that had already been decided for the majority of the country, including the rural inland areas.

The fragmentation of these inland areas into many small municipal systems with high investment requirements per capita and a long-term tradition of inefficient management and low tariffs virtually eliminated the viability of solutions with private sector participation. It was essential to merge systems upwards to a dimension that made it possible to manage them efficiently and profitably.

The sustainability of bulk systems of the inland areas was also at risk, because of the lack of sustainability of the small municipal systems themselves, as some of them collected less for the full water service from their customers than they had to pay to their bulk water company for treated water.

It was clear that there was no possible solution that didn't include the restructuring of municipal systems, but this required very difficult political decisions regarding the powers of the municipalities.

## 5.5.4 THE RESULTS

The annual results of ADP in 2004 were the best ever, but this meant very little in terms of meeting the anticipated corporate and especially national political objectives. It was soon realized that the solution to the internal ADP problems were within reach, but the overall success regarding the ambitious objectives presented above, namely (2) the universal population coverage, maintaining existing responsibilities of the municipalities for their systems; (4) promoting compatibility between bulk and municipal systems; and (5) promoting sustainable solutions and regional solidarity, depended on external factors that would require a difficult political negotiation between the government and municipalities.

It was technically clear that there was no possible sustainable solution for covering the whole country with conventional urban type solutions, maintaining the current excessive number of very small municipal systems, the lack of efficiency of many of them and their low tariffs, significantly distant from full cost recovery.

In the country, there was an earlier example of restructuring of municipal electricity distribution systems. Due to accumulation of debts of the municipalities to the then state owned electricity production company (later privatized), the municipalities were persuaded to concession their distribution systems. This type of solution was absolutely rejected by municipalities for their water utilities at the time, especially by the larger ones, who already had their own well established solutions. However, it was clear that the ADP restructuring solution and water utilities sector in the country depended in the first place on the sustainability of the municipal distribution systems.

A number of solutions for voluntary association of municipalities were analyzed and proposed at the time, but without success, because of political instability: there were three different ministers for the environment in one year.

Figure 5.5.1 was used at the time for communication purposes to present the problem to the government, municipalities and banks: in particular, although restructuring was centered on the bulk water companies, it was clear that any sustainable solution depended primarily on the restructuring of the municipal water companies. It was absolutely necessary to take early action to reduce the number of distribution systems; otherwise the whole system was at risk of collapse.

## 5.5.5 CONCLUSION

The leader of a restructuring process of water utilities must inevitably deal with the highest political levels, in a very complex, volatile and dynamic working environment. The initial objectives may be apparently clear, but they may be altered at short notice due to completely unforeseen reasons.

Restructuring a sector at a national level is, of course, much more complex than restructuring a single water utility, although the same basic principles continue to apply. The major differences are the level

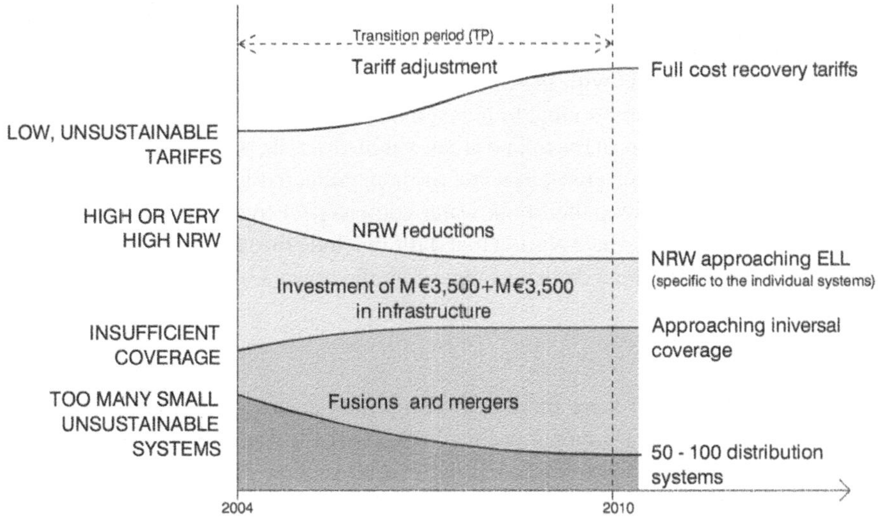


**Figure 5.5.1** A simplified restructuring model.

of political negotiation with many conflicting interests whilst the critical point is the definition of the sustainability boundaries of the systems. Where political conditions are able to be met, namely political support, time and stability, it falls upon the leader of the restructuring process to establish the necessary negotiations in order to move forward.

Obtaining more money, adjusting tariffs, reducing NRW and other inefficiencies are certainly difficult problems to resolve, but these are within reach for a motivated leader.

## 5.5.6 REFERENCE

*Águas de Portugal* (2005). Annual Report 2004 (Portuguese Version). *Águas de Portugal*, Lisbon, Portugal.

# Index

IWA Publishing's authorised EU representative for General Product Safety Regulations is Diane D'Arras, 15 rue Duret, 75116 Paris, France, e-mail: safety@iwap.co.uk.

Printed and bound by CPI Group (UK) Ltd, Croydon, CR0 4YY
16/06/2026
02140367-0003